T0259714

Textile Science and Clothing Technology

Series editor

Subramanian Senthilkannan Muthu, Hong Kong, Kowloon, Hong Kong

More information about this series at http://www.springer.com/series/13111

Subramanian Senthilkannan Muthu
Editor

Green Composites

Processing, Characterisation and Applications
for Textiles

 Springer

Editor
Subramanian Senthilkannan Muthu
Hong Kong, Kowloon, Hong Kong

ISSN 2197-9863   ISSN 2197-9871   (electronic)
Textile Science and Clothing Technology
ISBN 978-981-13-4712-2   ISBN 978-981-13-1972-3   (eBook)
https://doi.org/10.1007/978-981-13-1972-3

This Springer imprint is published by the registered company Springer Nature Singapore Pte Ltd.
The registered company address is: 152 Beach Road, #21-01/04 Gateway East, Singapore 189721,
Singapore

This book is dedicated to:
The lotus feet of my beloved Lord
Pazhaniandavar
My beloved late Father
My beloved Mother
My beloved Wife Karpagam and
Daughters—Anu and Karthika
My beloved Brother
Last but not least
To everyone working in the textile sector to
make it SUSTAINABLE

Contents

Natural Fiber-Based Hybrid Bio-composites: Processing, Characterization, and Applications

Rakesh Potluri

Abstract In this era, use of composite materials has become ubiquitous in aerospace, defense, automobiles, sports, and medical fields. But most of the commercially available composites are manufactured from the sources obtained from the fossil fuels that are depleting rapidly and are the major cause of environmental contamination. In addition to the pollution, the majority of these composites are nonbiodegradable and are ending up in the landfills. In order to deal with these sorts of problems effectively, there is an indispensable necessity for the development of sustainable bio-based composite materials, with eco-friendlier end of life possibilities. Natural fiber-based bio-composites have many distinct advantages with huge possibilities in different application areas which will have a great impact on the practical applications. This leads to a huge chunk of researchers to explore the performance of the natural fiber composites, investigating different methods for improving their characteristics. Hybridization is one such approach which can lead to the improvement in the mechanical, thermal, and dynamic characteristics of the composites. This chapter deals with the processing, characterization and different applications of the hybrid composite based on the natural fibers that were reviewed in the most recent context with mechanical properties as the main emphasis.

Keywords Hybrid composites · Bio-composites · Natural fiber composites
Natural fibers · Fiber extraction · Hybrid composite manufacturing
Property characterization · Applications

R. Potluri (✉)
Department of Mechanical Engineering, DVR & Dr. HS MIC College of Technology,
Kanchikacherla, Krishna 521180, India
e-mail: rakesh.potluri92@gmail.com; rakesh.potluri@mictech.ac.in

© Springer Nature Singapore Pte Ltd. 2019   1
S. S. Muthu (ed.), *Green Composites*, Textile Science and Clothing Technology,
https://doi.org/10.1007/978-981-13-1972-3_1

1 Introduction

Composite materials are a class of materials which are obtained by combining two or more materials which are dissimilar in nature at a microscopic scale, and possessing a distinct interface. This particular formulation leads to several distinct advantages that cannot be seen when these materials are used individually. Nowadays, we can see that there is a lot of increasing demand for composite materials used in different fields such as aerospace, defense, naval, construction, medical, sports industries, etc. This particular trend of increased usage of composites can be observed, especially in the automobile industry where the manufacturers are trying to increase the efficiency of the automobile by decreasing the weight of the components used in the automobile. A lot of researchers and scientists from around the world are putting effort and interest in the field of polymer-based composite materials as they offer several distinct advantages such as the high strength to weight ratios, high stiffness to weight ratios, tribological properties and their degree of flexibility for customization toward a specific application, their corrosion and chemical resistance properties, their ease of manufacturing, etc.

Traditionally, the polymer-based composite materials are reinforced with different forms of fibers such a roving, woven, braided forms, etc., which results in enhanced mechanical and thermal properties which can be tailor-made for the use in specific applications. The result of fiber being a reinforcement inside polymer is commonly referred to as fiber-reinforced polymer composite. Along with the fibers as reinforcements, some particulates can also be added to the matrix to tweak the change in different mechanical, thermal, and tribological properties of the polymer. The result of particulates being a reinforcement inside a polymer is commonly referred to as a particulate-reinforced polymer composite material. Some of the fibers that are used as the reinforcements can be either artificial/synthetic fibers or natural fibers also.

The increase in the consciousness regarding the depletion of the environment and decrease in the fossil fuel reserves has fueled a great need to look toward the development of new sustainable materials for the use in different fields for creating new products. This era has seen a very high boom in the research for developing new bio-composites as an effective replacement for composites that were traditionally created using synthetic polymers and fibers. There are several attractive features that are encouraging researchers to look toward natural fibers as a replacement for synthetic fibers. Some natural fibers can even exhibit similar stiffness properties that are comparable to that of the artificial fibers. One of the main advantages of the natural fibers is that they are abundantly available as an agricultural by-product or directly from natural growth itself. The natural fibers also require less amount of energy resources for processing; the process of extraction also is very flexible. Lower density is a very good feature of natural fibers. Fracture toughness, mechanical properties that are comparable to that of the artificial fibers, very minimal amount of health hazards associated with their processing and handling of these fibers, good machinability, improved surface finishing, biodegradability …etc., are the best features that are present with natural fiber-based polymer composites.

When natural fibers are used as reinforcement phase, these composites are referred to as Natural Fiber Composites (NFC's). In the present era, there is a constant research taking place towards replacing the synthetic fibers as reinforcements inside the composite with the natural fibers to make the composites more sustainable and biodegradable. Lot of researchers are also looking for replacing the synthetic matrix with that of the biodegradable natural polymeric matrix materials. This combination of both the matrix and the reinforcements that are obtained from the natural resources will give rise to the class of composites known as green composites or fully bio-composites.

This book chapter present's details about the topics of polymers, natural fibers, natural fiber extraction methods, production methods used for producing hybrid composites, property variations due to hybridization, applications, and future scope of these bio/natural fiber-based hybrid composites. In Sect. 1, an introduction to polymers and brief information about different types of natural polymers, an introduction and brief information about various natural fibers extracted from different parts of the plants and synthetic fibers is presented. Section 2 describes different types of hybridizations and its main purpose in terms of composite materials. Section 3 describes, in brief, different extraction and processing methods used for producing various natural fibers along with introduction and information on various processes for manufacturing hybrid composites. Section 4 describes information on how to characterize different properties of the fibers and composites along with the work done by different researchers on characterizing different hybrid composites focusing mainly on mechanical properties. Section 5 contains the information about the application of the natural fiber composites, hybrid composites made from natural fibers in various fields, and concludes by providing the present status and the future scope for these materials.

1.1 Polymers

Matrix phase is a very crucial material in the phenomena of load transfer from the surface of the composite onto the fiber reinforcement phase, environmental protection to the fibers from outside conditions, mechanical abrasion of the composite, chemical resistance, etc.. Polymers are one of the most abundantly used material as for the matrix phase in the composite materials. The applications of the polymers are abundant used in many different structural applications in various fields such as starting from aerospace, sports, naval, defense, construction, etc.

1.1.1 Synthetic Polymers

Polymers are classified into two types, namely, thermosets and thermoplastics that are formed by the combination of monomers linked together. Polymerization is the process of converting the monomers into a polymer. Synthetic polymers were created in the early twentieth century by chemical reactions. The applications of these polymers can be seen in many different fields. The different characteristics of the polymers are only restricted by the availability of the chemicals, thermodynamic laws, and the imagination, creativity of the chemical scientists or engineers. The polymers which are chemically derived by modifying the natural products like from the starch, rubber, lignocellulose, etc., are known as natural polymers. The properties of the polymers depend on factors such as monomer composition, molecular mass, branching structure, crystallinity, chain flexibility, etc., which can be tailored according to with the use of the processes such as blending, copolymerization, macromolecular architecture alternation, etc.

Thermoset Polymers Thermoset polymers are the polymers having a cross-linked chemical bonding between the monomers. The remelting of the polymer is prevented by the presence of this cross-linking type structure in the polymer on the application of temperature which makes it ideal for high-temperature applications such as aerospace, electronic applications. Thermoset polymers possess high mechanical strength, chemical and temperature resistance, structural integrity, and high stiffness. These are the major characteristics that are required for the industrial design of products. Thin and thick-walled structures can be created using these polymers. Cost-effectiveness, good aesthetic appearance, and flexibility in design are some of the other pros associated with thermosets. The major problems (or disadvantages) of the thermosets polymers are that they can not be recycled, remolded, and surface finishing is difficult. These thermosets also have lower impact strength or impact resistance. Various forms of reinforcements like fiber and particles or both can be added to these resins or polymers to form different composite materials. Epoxy, phenolic resin, polyester, polycarbonate, vinylester, silicon, urethane, polyimides, etc., are some of the examples for the thermoset polymers or resins.

Thermoplastic Polymers Thermoplastic polymers can be heated to make them flow from which we can say that thermoplastic polymers exhibit viscoelastic behavior. The curing process for a thermoplastic polymer is fully reversible, which leads to the remoldable and recyclable characteristics for this type of polymers. For this, remolding can be done without having any major effect on the physical properties of the polymers. These polymers offer high strength and shrink resistance. These are generally used for many structural applications which are subjected to lower forces and stresses. Chemical resistance and high impact resistance, higher recyclability, and eco-friendly manufacturing can be possible with these types of polymers. Some of the major drawbacks with these type of polymers are that they are expensive when compared with thermosets and has lower temperature resistance. Polypropylene (PP), polyvinyl chloride (PVC), polystyrene, polyether ether ketone (PEEK), high-density polyethylene (HDPE), low-density polyethylene (LDPE), and polyethylene terephthalate are some of the examples of thermoplastic polymers.

Different processing methods such as thermoforming, Resin Transfer Molding (RTM), casting, pultrusion, spray up, filament winding, and compression molding are some of the techniques used for the processing of thermoset polymers. Extrusion, injection molding, blow molding, rotational molding, and screw extrusion are some of the techniques used for processing thermoplastic composites.

1.1.2 Natural Polymers

Natural polymers are more sustainable and biodegradable in comparison with that of synthetic polymers. Natural polymers can be classified into three types: (1) Natural polymers obtained from sources such as starch, protein, and cellulose sources. (2) Polymers obtained through microbial fermentation such as polyhydroxyalkanoates (PHA's). (3) Synthetic polymers obtained from natural monomers such as polylactic acid (PLA). When the biodegradable material is obtained fully from the naturally available sources, they are known as green materials or green/bio polymers. Tensile strength and modulus of elasticity comparison for several natural polymer are illustrated in the Fig. 1. From Fig. 1, it can be noted that polyglycolic acid has the highest strength and modulus values, which makes it a desirable matrix material for medical applications.

Polylactic acid (PLA) Polylactic acid (PLA) is a very good polymer, which is sourced from renewable resources. It is a biodegradable and a versatile polyester which is mainly based on corn, wheat, beat, and other starch-based products. It has a higher tensile strength and elastic modulus in comparison with that of the polypropylene (PP), Polystyrene (PS), and polyethylene (PE); it is brittle in nature and has lower toughness. PLA can bear only less than 10% elongation before breakage. PLA is produced by the process of fermentation of lactic acid produced from natural sources. The major disadvantages such as low heat resistance coupled

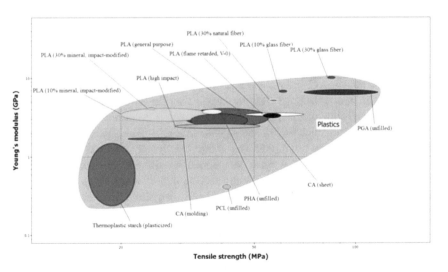

Fig. 1 Young's modulus versus tensile strength of natural polymers [1]

with its high cost, low aging resistance for the polymer material with lower impact strength, limits its applications in general composite materials.

Polyhydroxyalkanoates (PHA's) PHA's are the biodegradable polymers. They can be originated naturally by the process of bacterial fermentation. These polymers are classified into two groups depending on the number of carbon atoms in the PHA's monomers. One of the types is the polymer containing 3–5 number of carbon atoms which is known as the short chain length PHA's. The second type of polymer containing 6–14 number of carbon atoms is known as the medium chain length PHA's. Several types of microorganisms can be used for easy degradation of these types of polymers.

Poly (3-hydroxy burate) (P(3HB)) is an example of short chain length PHA's. The toughness and processing of PHB can be effectively improved by adding [3-hydroxy valerate (3HV)] to the fermentation process used for producing these types of polymers. Some of the disadvantages associated with the PHB polymer are the poor mechanical properties, high fragility, and processing difficulty in the molten state due to the degradation of the properties of the polymer which starts at around 200 °C only.

Aliphatic Polymers These are obtained by the process of the polycondensation reaction, synthetically. Glycols and aliphatic dicarboxylic acid are used for the process of polycondensation. These are generally sourced from petroleum, but 1,3-propanediol and succinic acids are obtained naturally. Due to the low polymerization and failing to produce high molecular weight polymers and block polymers, this is not an effective method. The oligomers are the output of this method which has very poor thermomechanical properties. The oligomers produced through this method can be processed through a sequential chain extension method which includes reacting the oligomers with sebacoyl chloride to produce high molecular weight polymers. Another synthetic way to produce high molecular weight polymers is through ring-opening polymerization of lactones and cyclic diesters. Aliphatic polyesters can also be synthesized by bacterial fermentation of sugars and lipids to form polyhydroxyalkanoates

Aliphatic–Aromatic Copolymers The aliphatic–aromatic polyesters can be combined with aromatic polyesters by copolyesterification of both monomers to yield a polymer with mechanical and biodegradable properties. The reaction of aliphatic and aromatic dicarboxylic acids with aliphatic glycols using the process of polycondensation forms an aliphatic–aromatic copolymer. Best example for these types of polymers is the Ecoflex developed by BASF, Ecoflex, which has mechanical and thermal properties equivalent to that of low-density polyethylene (LDPE). This material, when formed into sheets, can allow the water vapor to be permeable.

Polyester Amides Polyester amides (PEA's) are the result of combining ester and amide linkage in a chain. They are biodegradable and possess the properties of both polyamide and polyester families. Polyesters degrade from the ester linkage cleavage and are better soluble in organic solvents and possess good mechanical properties. Polyamides possess higher thermomechanical properties, due to the fact that hydrogen bond is absent in the linkage. Polyamides also do not degrade inside the human body which makes it a good choice for biomedical applications. Combining both esters and amides results in good mechanical, thermal, and biodegradable properties.

Polybutylene Succinates Polybutylene succinates (PBS) are a well-known biodegradable polymers coming from the family of polyesters. These are produced using butanediol, succinic acid or other carboxylates and alkyldiols. The properties of the PBS are similar to polyethylene. It is thermally stable up to 200 °C with good mechanical and biodegradable properties. It is industrially produced through the process of condensation polymerization of succinic acid and butanediol. This polymer can be processed through extrusion and blow molding and injection molding process. It is mostly used for packaging films, bags, and hygiene products.

Polyvinyl Alcohol PVA is a water-soluble resin and translucent in nature. It is mostly used for coating papers and fabric sizing. The PVA is available in two forms of either completely hydrolyzed and partially hydrolyzed forms. Completely hydrolyzed PVA is highly soluble in water and slightly soluble in organic solvents and it is vice versa for the partially hydrolyzed PVA. PVA is produced by combining polyvinyl acetate and methanol in the presence of alkaline catalysts such as sodium hydroxide. PVA, due to the presence of OH group in the linkage, exhibits a high rate of biodegradation. It is mainly used in the coating of papers to make it resistant to greases and oils. It is also used as the coating for yarns to improve the strength and reduce the absorptivity of the fiber.

1.2 Natural Fibers

Fibers are generally bifurcated into two types: synthetic fibers and natural fibers. Sustainability, renewability, abundance, and biodegradability are the favorable features of the natural fibers. Natural fibers also have the potential to address the problem of waste management, and has the ability to create local micro-economies in the developing countries. The strength and mechanical properties of the natural fibers have encouraged materials scientists to bring a lot of natural fibers as reinforcements into the composite materials. The classification of the natural and synthetic fibers is shown in Fig. 2. The major sources of the synthetic fibers and the energy required for processing these fibers are sourced from petrochemicals. The availability of the petrochemical resources which are rapidly depleting and their reserves are uncertain in nature. Moreover, using these petrochemicals yields to a lot of pollution and damages the environment. To counter all these factors, scientists had taken up the task of replacing these synthetic fibers with the natural fibers. This has led to the research on natural fiber-based composites or bio-composites.

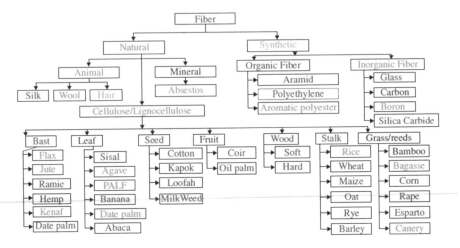

Fig. 2 Classification of fibers [1]

Natural fibers can be sourced from different sources such as animal, plant, and mineral fibers. Wool, silk, and hair are the examples of fibers that are sourced from the animals. Fibers can be extracted from several parts of the plant like its stem/bast, leaf, seed, fruit, etc. The classification and examples for several types of natural fibers are provided in Fig. 2. Animal and mineral-based fibers have not been used in the production of composites but rather used for producing clothes, paper, handi-crafts, etc. A new mineral-based fiber known as the basalt fiber which has very high mechanical, chemical, and thermal properties came into the picture. The researchers are trying to use this fiber as an effective alternative to synthetic fibers. The plant fibers are also subdivided into wood-based and non-wood based natural fibers based on their origin. A lot of the applications of the natural fibers comes to a good use as the bio-composites in the automotive, construction, sports, and marine industries. If observed clearly the use of the natural fibers as a reinforcement dates back to 300 B.C. where straw was used as a reinforcement inside the clay to prepare a composite brick. Composite bows were used by Mongolians which were made from horns is another example for the use of the natural composite by humans. The classification of natural fibers sourced from different sources such as leaves, bast, minerals, etc., are represented in Fig. 2. The comparison of the properties such as specific Young's modulus and tensile strength of some natural fibers with some metals and polymers is represented in the Fig. 3. Different natural fibers extracted from different plants are represented in Fig. 4 for a generalized sense of how the fibers may look like after extraction.

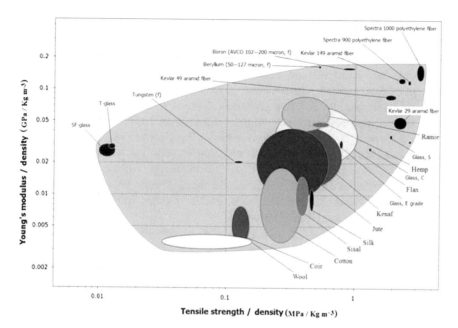

Fig. 3 Specific tensile strength versus specific young's modulus of natural fibers, metals, glass and polymer fibers [1]

Fig. 4 Different natural fibers extracted from different sources [2]

1.2.1 Wood-Based Fibers

Wood is an example of the natural composite material. It is made up of lignin, cellulose, hemicellulose, and other additional materials. These other materials consist of fats, protein, inorganic salts, flavonoids, alkaloids, waxes, terpenes, glycosides, simple and complex phenolics, lignans, stilbenes, pectin, starch, tannins, mucilages, gums, saponins, simple sugars crucial oils, proteins, and many more compounds. This is known as extractives. They tend to weigh around 20% in the total dry weight of the wood. Even being in very low quantities also, these extractives play a major role in final mechanical strength and quality of the wood but these extractives also tend to decrease the thermal properties of the wood. Some researchers observed that removing these extractives tend to improve the thermal properties of the wood-based composites.

Wood can be mainly divided into two types, namely the hard and the softwood. If the wood is sourced from the gymnosperm tree, it is softwood and the wood sourced from the angiosperm tree is the hardwood. The woods are made up of cells that are connected in different fashions. Two types of cell arrangements can be observed inside the wood. One type of arrangement is the axial cells which run parallel to the long axis of the wood plant. The second type of arrangement is a radial cell which is aligned perpendicular to the long axis of the wood plant. The softwoods are made up of axial tracheids which are usually 1–10 mm in length and have an aspect ratio equivalent to that of 100. These axial tracheids are the structures which provide the required strength for the softwoods.

The hardwood cell structure is complex in comparison with the softwoods. Hardwoods consist of axial fibers and axial parenchyma, which are usually 0.2–1.2 mm in length and having a width of equivalent to that of the axial tracheids, those which are present in the softwood. The mechanical strength and the density is determined along the thickness direction. The thicker cells yield stronger and dense structures for the wood. Generally, two types of fiber forms can be obtained from both the soft and hardwoods. Single fibers which have higher aspect ratios and small fibers with lower aspect ratios. The single fiber is known as wood fibers and short fiber bundles are extracted as wood flour. The major sources to obtain the sawdust, shavings after cutting wood, and woodchips are the sawmills and woodworking centers which are postindustrial sources. Generally, the wood flour is one of the sources used as filler materials inside composites rather than as reinforcement materials.

1.2.2 Bast Fibers

These are derived from the bast of the plant. They are derived from the inner bark/phloem of different plants. The woody xylem core is normally covered by the phloem which is one of the sources of strength and stiffness to the plant stem. The discussion about some of the bast fibers is provided in the following paragraphs.

Flax: Flax is a plant that is being cultivated for the purpose of extraction of fiber and seed oil. This is one of the old natural fibers that is widely used in the manufacturing of modern textiles. It is grown in the countries like India, Netherlands, France, Spain, Russia, etc., here, there is a temperate climate exists. 75% out of total length can be utilized as the fiber in the total length of the plant.

Hemp: Hemp is a plant with a number of inherent advantages. It usually grows in central Asia and central Europe. It has a very high growth rate combined with good resistance to pests and viruses which leads to very fewer requirements for pesticides, herbicides, fertilizers, and fungicides. Its root structure helps to maintain the soil structure, and has the capability of restoring nutrients to the soil. It has a higher tensile strength and lower moisture absorption and elongation in comparison with the flax fiber.

Nettle: Nettle fiber is one of the very ancient fibers. It was cultivated extensively for the fiber during the world war 2. Its production has decreased as it has a very long growth cycle of more than a year, even if it is possessing a higher tensile strength and modulus values in comparison with other fibers that are sourced from the bast of the plant.

Jute: Jute is one of the most common fibers that grow in the Mediterranean, and in Asian countries like India, Indonesia, Nepal, Thailand, Bangladesh, and Brazil. It is a plant which is grown only for the fibers. Due to high lignin content present in the fiber, it has lower elongation and tensile strength, lower chemical, and moisture resistance in comparison with the hemp and flax fibers. The major advantage of the jute fiber lies in the fact that this fiber is fire and heat resistant.

Kenaf: A plant that is mostly native to central Africa and subtropical Asia. It grows very rapidly and contains long and short fibers, which can be extracted from the same bast. It is very rough in texture and brittle in nature which makes it difficult to processes this fiber. The mechanical properties of the kenaf fiber is very similar to that of the jute fiber.

Ramie: It is one of the oldest crops that is being grown in Asia, which can be harvested several times, around four to six times, a year. The ramie fiber has high specific strength and modulus and has mechanical properties comparable to glass fiber with high chemical resistance as well.

Kudzu: It a fiber from Japan and China and is a plant which has expanded rapidly through the seeds and stolons. The fiber extracted is generally used for basketry also in the clothing industry. The rapid expansion of this plant leads to depletion of the quality of the soil and its expansion has to be controlled.

Okra: This plant was originally from Africa. It is usually grown in the tropical, semitropical, and warm climates. It is a perennial plant which is most often grown not for its fiber, but for the okra vegetable. More concentration is now being given to the fibers that are obtained from the bast of the plant after harvesting the required vegetable. It has comparable properties with other natural bast fibers like jute, kenaf, etc.

Roselle: This is a plant that takes a long time for its growth, which is a major drawback of this plant. It is grown in India and Africa. China and Thailand are the largest exporters of this fiber which is a very good replacement for the jute fiber.

1.2.3 Leaf Fibers

Leaf fibers are the group of multicelled lignocellulose fibers sourced from the plant leaves. The leaves that are suitable for fiber extraction should be long, linear in structure. Generally, the fibers are extracted from the leaves through mechanical means. The discussion about some of the leaf's fibers is provided in the following paragraphs.

Sisal: The sisal majorly grows in the tropical climate. Brazil and Tanzania are the major producers of this fiber. It has a very high lifespan of 7–10 years. It has three types of fibers which are the structural, arch, and the xylem fibers. The structural fibers are important for the commercial use of the sisal fibers but the mechanical strength is the highest for the arch fibers and is used for the short fiber applications; xylem fibers are lost during the extraction phase of fibers from the leaves of the sisal plants. High strength, durability, resistance to salt-dependent degradation, ability to be dyed, and stretching are some of the favorable characteristics of sisal fibers, but all the favorable characteristics tend to deteriorate with the increase in temperature.

Abaca: Abaca is a fiber extracted from the leaves of the specific species of the banana plant, which is native to the Philippines. This fiber is generally used for producing ropes and twines. It is a hard fiber and is the strongest fiber that can be sourced from the plants. It has an excellent resistance to sea water degradation.

Phormium tenax: Phormium tenax is a perennial plant which is most commonly cultivated in New Zealand. These plant leaves are one of the important sources for the natural leaf based fibers. The fiber extracted from the leaves of this plants are majorly utilized to prepare the objects such as mats and ropes due to its properties rather than a reinforcement inside a composite material.

Henequen: Henequen is a plant that is related to the sisal fiber family and it has lower properties in comparison with that of the sisal fibers. It is grown in Africa, South and Central America, and Asia. It has low mechanical properties and salt degradation in comparison with those of other leaf fibers. The toughness and resilience are the important properties exhibited by the henequen fiber, which are important properties for some composites.

Pineapple Leaf Fiber (PALF): Pineapple leaf fiber is obtained as a by-product coming from the pineapple fruit cultivation, which is majorly grown in tropical climates. The mechanical properties of this fiber are very similar to that of the jute fiber. It is highly hygroscopic in nature. It is commonly combined with silk for creating the fabric for the textile industry.

Banana: Banana fiber is extracted from the banana truck which is a circular leaf structure often mistaken to be as bast. It is useful for preparing the biodegradable products in Asia and throughout the world. It has been used as a textile fiber from ancient times. The cellulose and structure of the banana fibers are similar to the sisal fiber, but the lower spiral angle in the structure of the fiber results in a higher tensile characteristic in comparison with that of sisal fiber.

Curaua: Curaua is a plant which grows in regions with semiarid climates. It is a hard fiber whose chemical composition is similar to that of several other plant fibers. Due to the mechanical properties of this fiber, it has been used for automobile applications.

Date palm: Data palm trees are mainly cultivated for the fruit known as the dates, which are usually found in the Middle East, Africa, America, India, and Pakistan. The hard shell found on the tree is from which the fibers are extracted that can be used to prepare ropes and basketry.

Piassava: Piassava fibers are extracted from the native palm trees. The tensile properties of these fibers are more in comparison with the coir and lower in comparison with the sisal and jute fibers. It was mainly used for industrial mats and brooms.

1.2.4 Grass Fibers

Grasses are another good source of lignocellulose fibers, which can be derived from a variety of grasses. They are good due to their low cost, simple harvesting, and processing requirements. But the properties of grasses are lower in comparison with that of bast and leaf fibers, which makes them suitable to be used as short fiber reinforcements. Natural fibers can be extracted from a variety of grasses such as the bamboo, wild cane, Indian cane, switchgrass, alfa, as well as different types of straws, etc. Grass fibers usage was investigated and are being used in the automotive industry in European countries. America is also advising its automotive giants to use these natural fiber composites in manufacturing different components that are used in their automobiles as well. The fibers from these grasses, are extracted either by retting or mechanical decortication.

Straw: Straws are leftovers obtained from the cereal stalks like rye, rice, and wheat plants. They are obtained from the plants after the grains are removed from the rye, rice, and wheat plants. They are generally used as the fodder for animals and to produce biofuel, bedding, thatching purposes. The mechanical properties of the fibers extracted from these straws are lower in comparison with bast and leaf fibers.

Bamboo: Bamboo is one of the largest plants in the grass family, with a very high growth rate and low growth cycle. Fiber can be extracted from the pulp of the bamboo plant. It mainly belongs to Asia. This is usually used as a traditional food source, for construction purposes, making furniture, paper, and textiles. These bamboo fibers are the ones, which have good properties making them attractive materials for reinforcing the composite materials. A lot of research efforts are being focused in many parts of Asia and throughout the world on effective usage of the bamboo fibers as reinforcement phase in the composite materials.

Alfa: Alfa is a perennial grass found in the southern region of Spain and northwest region of Africa. The main application of this grass is for pulp making used in the production of the paper. But, the fibers extracted from this alfa grass are short fibers having high mechanical properties. That is the reason, a great attention is given for the usage of this alfa fiber for the preparation of composite materials in recent times.

Wild cane: Wild cane grass is one of the most abundantly available natural resources of grass. The density and mechanical properties of the fibers extracted from the wild cane grass have comparable properties to that of the sisal and banana fibers.

1.2.5 Fruit and Seed Fibers

These are the fibers that are derived from either the fruit of the plant or seeds of the plant. Coir and oil palm are examples of fibers extracted from the fruits of a tree. Cotton is the best example of the fibers extracted from seeds. Cotton fibers form a protective layer around the seeds. Some of the fibers extracted from fruit and seeds of the plants are discussed in the following paragraphs.

Oil palm: It is a native plant to the Congo basin and to West Africa. It yields different products such as palm fruit, medicine, fuel from the trunk, palm oil, etc. These fibers are extracted from the waste using the process of retting. The major disadvantages of the palm fibers are that the fiber diameters vary significantly, which influences the tensile properties of the fibers, whose properties are lower in comparison with the other natural fibers.

Coir: Coir fiber is a fiber extracted from the outer shell of the coconut fruit, which is abundantly found in the subtropical and tropical regions. Coir which is extracted before the coconut that is ripened is known as white coir. The coir which is extracted after ripening of the coconut is known as brown coir. The coir has a relatively low mechanical strength and modulus, which leads to the fact that it is not a good reinforcing material. But the major advantages of coir lie in the fact that it has low thermal conductivity, density, and higher elongation. The coir fiber tends to be used as a rope in most of the countries especially in Asian countries.

Cotton: These plants are usually grown in the tropical and subtropical regions of the world and is the leading fiber crop that is being cultivated across the world. Cotton fibers possess lower mechanical properties in comparison with natural fibers especially the natural fibers which are obtained from the bast part of the plant. But the cotton is an abundant fiber used in the modern textile industry. It is also mixed with other natural fibers for the purpose of manufacturing stronger textiles.

1.2.6 Husk/Hull

The by-product that is obtained after removing the shell from the grains like soybean seeds, wheat grains, rye, rice, palm, kernels, sunflower seeds, etc. Seeds are the main products that will be obtained from the considered plants and husk/hull can be considered as an agricultural by-product. The husk is separated from the grains using the process of mechanical dehulling. This husk/hull is separated from the seed, and then usually sent to be used as animal feed due to the presence of high amount of fiber content. To address the problem of reducing this husk/hull waste streams, researchers have focused their attention toward the usage of these hulls in the polymer composite materials. The availability of the unprocessed hulls is not that high, and lack of studies on the mechanical properties of this husk/hull and their shape and structure is a major problem that is hindering the usage of these as a reinforcement in composite materials.

Sugarcane bagasse is one another by-product obtained after crushing of the sugarcane. This bagasse is generally used for as the cattle feed, a source for producing biofuel, and a source for producing paper. The fibers extracted from the bagasse are usually around 1.2 mm in length and with an aspect ratio of around 80 and it has similar performance as that of the wood flour and is a candidate for short fiber reinforced composites. Corncobs are one of the sources for the use in the biofuel industries. Starch can be extracted from the corn and other starchy plants which can be processed to produce biodegradable polymers.

1.2.7 Protein Fibers

Natural fiber can also be sourced from various protein sources in addition to the lingo cellulose sources. For the protein-based fibers, the amino acid protein structure is the basic building block. These structures are based on the sequence of amino acids connected by peptide bonds. For the purpose of reinforcements, only a few types of proteins known to be as sclero proteins which are in the form of long protein filaments will be considered. Keratin, elastin, Fibroin, and collagen are some of the fibrous proteins. These are most abundantly found in different parts of the vertebrates which makes them not a good source of protein fibers. Two types of keratin namely alfa keratin and beta keratin are generally found in nature. Alfa keratin is the primary source found in the hairs, horns, claws of the mammals, while the beta keratin is a source that is found in the structure of the shells, scales on the skins of the reptiles and scales of the birds, etc.

Feathers: Feathers are a by-product obtained from the poultry industry. Feathers are made up completely with beta keratin proteins. Feathers typically have a length of 5–1700 µm and 10–40 µm in diameter. These fibers have high acoustic and thermal insulation characteristics due to the presence of inherent air gaps in the keratin microfibrils. Along with their lower density, these feather fibers also possess good mechanical properties and modulus values as well.

Wool: Wool can be commonly harvested from the animals like sheeps, camels, goats, and alpaca, which is being widely used for producing clothes and fibers for the modern textile industry and is one of the most commonly found protein fibers. The research on how to harvest more wool and use it as fiber reinforcement phase in the composite materials has increased in recent times and many researchers are currently working on this particular area. But, the major problem lies in the fact that these fibers are costly, has weak mechanical properties, their nature is very similar to that of the cross-linked rubber, and finally the variations in lengths and characteristics of the fibers, which makes it very difficult to use these fibers as a potential reinforcement phase in the composite materials.

Fibroin: Fibroin is a form of protein which is created during the production of silk. Fibroin is a protein which is responsible for inducing the good tensile strength and modulus and performance. Silks are produced by many different types of arthropods, but the most common source of silk is from the bombyx mori silkworm larvae, which forms a cocoon made out of silk fiber. Silkworm silk has a good amount of tensile and Young's modulus values with the ability to form uniform continuous fibers. But the major challenge occurs in the fact that this fiber is rather expensive. The production of this fiber is also very low and this silk extracted from the bombyx mori silkworm is used for producing high-end textiles. Another well-known source of silk is the spider silk which holds good performance characteristics. But the production of this fiber is very costly and a difficult affair. But, seeing its potential for usage as a reinforcement in the composite materials, researchers have developed a way to synthetically produce this fiber, but the process is also very expensive and time-consuming and needs future improvements so that it could be applied at a commercial scale.

Different factors affect the properties of the natural fiber at different stages of its growth and extraction. The fiber properties and performance are based on the factors such as plant growth, harvesting stage, fiber extraction methods, and supply chain. In plant growth stage, the major determining factors are the plant species, how the crop is cultivated, the location of the crop, part of the plant from which the fiber is extracted, and the local environmental conditions. In the harvesting stage, the development of the fiber which can be determined by the fiber ripeness plays a crucial role. The fiber ripeness affects the fiber performance through its effect on the fiber surface coarseness, thickness of the cell wall, the bonding between the fibers, and the surrounding matter in the structure. Fiber extraction and treatments play a major role in determining the properties and performance of the natural fiber. Different extraction methods like mechanical decortication, retting, and dehulling have a profound effect on the physical and chemical properties of the natural fibers. Some chemical treatments are performed during the extraction stage of the fiber which also affects the fiber surface causing a change in the properties of the fiber. During the supply stage, the conditions in which the fibers are being transported, fiber storage conditions, and duration of storage which degrades the properties of the fibers as long durations increase the fibers age.

Different factors affect the performance of the natural fiber composites which includes the fiber section and fiber source, appropriate matrix selection, interfacial strength or bonding, the orientation of the fiber in the matrix, the distribution of the fiber in the matrix material, adopted manufacturing process for preparing the composite and manufacturing defects and environmental conditions.

1.2.8 Synthetic Fibers

These are the fibers which are artificially derived from the polymers which are not available naturally. They are mostly derived from the petroleum-based by-products and are developed in the laboratories and at industrial scale. These are derived from the combination of different chemicals, they are generally very long and possess high mechanical, thermal, and physical properties. These properties can also be greatly varied by using the different combination of chemicals and processes that are used to create these synthetic fibers. nylon, polyesters, acrylics, polyurethanes, etc., are some of the so-called polymer forms from which these types of fibers can be developed. Kevlar or aramid, carbon, and glass fibers are the three most commonly used synthetic fibers in composites in many different fields of aerospace, defense, construction, sports, naval, etc. Figure 5 shows the images of the glass, Kevlar, and carbon fibers.

Kevlar fiber: Kevlar fibers are also known as aramid fibers. Para-phenylenediamine is combined with terephthaloyl chloride, which forms the aromatic polyamide threads. These threads are nothing but the Kevlar fiber strands. This is a very expensive process that is used for manufacturing these fibers, but very high-performance fibers with very good impact resistance can be obtained. Different types of fibers such as Kevlar, Kevlar49, and Kevlar29 exists. Generally, these fibers are used for producing the body armors.

Glass fibers: Glass fibers are most commonly used synthetic fibers in the composite industry for many several applications. These are most versatile and cheap fibers in comparison with that of the Kevlar and carbon fibers. These fibers are manufactured through the process of drawing the liquified glass with 50% silica and with other mineral oxides. These fibers are cheap to produce and possess desirable properties such as lightweight, high strength, less brittle, and stiffness which are beneficial. Different grades of glass fiber such as E-glass, S glass, A glass C glass, etc., exists. These are mainly used in automotive body building, bulletproof glasses, fuel tanks, and many other components in the aviation, sports, construction industry, energy sector, etc.

Carbon fibers: Carbon fibers possess very high specific strengths and stiffnesses. They also possess extremely high thermal and chemical resistance and properties in addition to its strengths. Polyacrylonitrile (PAN) is the main source from which, about 90% of the carbon fiber is being currently produced and the

(a)   (b)   (c)

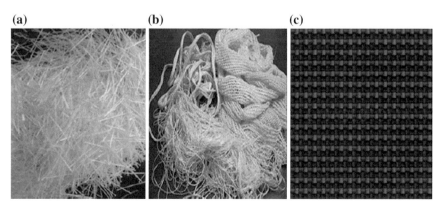

Fig. 5 **a** Glass fiber, **b** Kevlar fiber, **c** carbon fiber [2]

remaining comes from the rayon and the petroleum pitch. PAN and rayon are organic polymers which contain long strings of molecules, which are bonded together by carbon atoms. The cost of carbon fiber is very high which limits its usage in the composite materials in other fields except for the aerospace applications. Different forms of carbon fibers such as T800, carbon, etc., exists. A lot of research is also being conducted on how to extract carbon from natural sources for producing carbon fibers and also how to decrease the manufacturing costs of these carbon fibers. These fibers are used in manufacturing satellites, rockets, aerostructures, high-end luxury and sports cars.

Basalt fiber: Basalt fiber is a mineral fiber composed of similar minerals to that of the glass fibers. It has very good thermomechanical, physical properties in comparison with that of the glass fiber. It is produced by melting of the by-product of basalt rocks obtained from quarries. It is extruded through a small nozzle at a temp of 1500 °C and yield a fiber which is typically around 20 μm in diameter. It has comparable properties to that of carbon fibers. It has very high thermal, chemical resistance. It has all the characteristics similar to that of synthetic fiber but is obtained from a natural means. Some researchers have dealt with the basalt fibers treating it as a synthetic fiber and others have done the opposite by considering as a natural fiber.

2 Hybridization

Hybridization is the process of inclusion of the more than one type of reinforcements in a matrix material in terms of composite laminas and stacking up plies made from different types of composite material laminas to construct a laminate in terms of laminates. Hybridization can be seen as an effective way of reducing the disadvantages of one type of reinforcements by another one. Hybridization when coming to the natural fiber-based composites, most of the times the natural fibers are being hybridized with the glass fiber in a matrix material which may be either synthetic or bio-based polymer.

The major issue with this type of hybridization of natural fibers with glass fiber or any synthetic fibers, in general, is that the main advantages of the natural fiber composites like biodegradability, easy handling, and processing, reduced health hazards, sourced from the natural sources are being reduced. But on the other side, this type of hybridization can reduce the cost, weight, and the environmental issues created by the artificial fiber composites. On the other hand, the different natural fibers can be combined to prepare hybrid composites, and which also can be exhibiting the positive hybridization effects. Hybridization of the resins is also another concept and research has shown that this also has an enhanced performance on the composite material, but most of the research is being put on the hybridization of composites with different fibers.

3 Processing of Natural Fibers and Hybrid Composites

3.1 Fiber Extraction Methods

Natural fibers are usually extracted through different methods which may be manual or mechanical in nature. Most of the natural fibers are extracted through the process of either retting or decortication. Generally, the combination of both of the process may also be used. In order to remove hull/husk fibers, a process of mechanical dehulling may be applied. The information regarding the various retting processes, mechanical dehulling, and decortication is provided in the following subsections.

3.1.1 Retting Process

Retting is an effective method for extraction of the bast and leaf fibers. This method has a great effect on the final quality of the fibers produced using this method. Generally, the retting process can be accomplished in four different ways, namely, the biological retting, mechanical retting, chemical retting, and physical retting.

Biological retting process: Biological retting can be again classified into two types, namely, natural retting and artificial retting. Natural retting can be again performed in two ways, namely, dew/field retting process and cold-water retting process.

Cold-water retting process: Cold-water retting breaks down the pectin present in the plant bast bundles in the water with the help of anaerobic bacteria. The outer hard layer of the plant is destroyed by the water that penetrates through the central stalk portion and swells the inner cells to allow for increased absorption of moisture and bacteria. Depending on multiple factors such as water type, temperature, bacterial resistance, etc., the retting process may take about 1–2 weeks to complete. Good quality of fibers can be extracted by employing this method. But the major cons of this method are the time consumption and the release of unnecessary pollution into the environment through the disposal of wastewater which is contaminated with the organic fermentation by the anaerobic bacteria. To reduce the time of the retting process, warm water retting can be performed. The stems, leaves, and bast are soaked in hot water which is usually maintained at a temperature of between 25–40 °C, rather than in the cold water. This accelerates the process of retting and brings downs the retting time from 2 weeks to 3–5 days.

Dew/field retting process: Dew/field retting is a common procedure that is undertaken for the extraction of fibers in the areas with very limited water sources. It takes the help of the dew for moisture instead of water and microorganisms to break the fiber free from the bast/stem or leaves. The duration of this type of retting process is very long and takes somewhere about 3 weeks to 6 weeks. This retting is very much dependent on the factors such as the climate, bacteria generation, and the humidity in the climate. The fibers extracted through this retting procedure are darker in color and are of poor quality in comparison with the fibers extracted through water retting procedure. The major disadvantage with this type of dew retting is mainly the unpredictability of the retting process, high amount of time requirement, discoloration of the fibers, pollution caused by the fermentation to the environment, and low quality of fibers obtained. In order to increase the fiber quality, a thermal-assisted modified dew retting process is devised, which makes use of temperature application for the existing retting process. But these extra step requirements increase the cost of extraction of the fibers. Artificial retting is nothing but the introduction of temperature controlling, bacterial control, and other process control into retting process, through artificial sources in order to accelerate the process of retting.

Mechanical retting process: Mechanical retting is also known as green retting. It is a simple as well as a cost-effective method that can be applied for separating the fiber from the xylem and the plant straw. Even this procedure requires drying of the stalks or leaves; much of the dependence on the weather and bacteria are eliminated from this process of retting. The dried stalks are used for the fiber extraction using the simple mechanical machines to blast off the outer skin and to get the fibers, instead of the bacteria and water doing this process as in the biological retting processes. The fibers obtained by using this particular process of retting are coarse in comparison with those fibers obtained by biological retting procedure. This is the main reason which limits the usefulness of fibers extracted through the process of mechanical retting to be used as reinforcements in the composite material

Physical retting process: In order to extract clean and fine fibers with high and consistent quality, physical method of retting may be applied. This is also known as wet retting process. The major advantage of this type of retting process is that it can be used for modifying the characteristics of the extracted fibers for different applications, through the adjustment of different processing parameters. Three types of methods are included in the physical retting process, they are the ultrasonic retting, enzyme retting, and the stem explosion methods (STEX).

The ultrasound retting is a very good replacement for avoiding the use of the unreliable dew retting step in both the biological and the mechanical retting procedures. Ultrasound is applied for the process of extraction of the fiber from the stalks of the plant. The fibers obtained through this procedure are usually of non-textile in grade. The STEX procedure of retting can be applied for extraction of the fibers with high finesses and properties that are comparable to that of the cotton fibers. This is one reason, which makes it suitable for applying the fibers extracted through STEX retting process in the textile industry. The enzyme retting process is the costliest process of all the other retting processes. It takes about 2–24 h to produce the fibers using this retting procedure. This retting procedure is able to produce long fibers and undamaged individual fibers with a good inherent fibers strength.

Chemical retting process: Chemical and surfactant retting methods are used for extraction of the fibers using the application of warm or heated water with the addition of different types of chemical to it. This is also like the enzyme retting process, which is a very time effective retting procedure. Different chemical modifications are possible to dissolve the pectin present in the stalks and to separate the fibers. Different chemicals such as sulfuric acid, chlorinated lime, potassium hydroxide, sodium hydroxide, and sodium carbonate can be used for the chemical retting process. The major disadvantage with this type of retting is that it is very expensive to perform. Fibers with consistent and good properties can be obtained by applying this retting procedure.

The procedure through which the fiber is extracted can be one of the very crucial factors that dictate the fiber morphology, surface composition, mechanical, and thermal characteristics of the fibers. Methods such as physical retting, chemical retting, and optimized biological retting have shown results for better and easier separation of the fibers from the stalk/bast of the plant while having a minimum mechanical loading effect on the fibers extracted. Sisal, Roselle, agave Americana, seagrass, ripe coconut coir, coconut leaf sheaths, curaua, vakka, palm fiber, bamboo fiber, okra, elephant grass, etc., fibers are some of the examples of the fibers that are extracted through various retting processes, and then dried to remove moisture from the fibers and make them useful for reinforcement inside a composite.

3.1.2 Mechanical Decortication

Most of the leaf fibers are extracted using the process of mechanical decortication. There are a variety of diversified plants, whose leaves are applicable for extraction of the leaf fiber after harvesting the crops. First, to help with the easy removal of the fibers from the leaves during the process of decortication and extraction, water retting can be done for the leaves for a period of about 1–5 days. This is not the case with all of the leaf fibers. The leaves of the plants are crushed under pressure with the help of rotating drum consisting of blunt blades. This crushing helps to break the fiber loose from the other hard parts, components, and structure of the plant leaf. After decortication of the leaf, the fibers are washed, and then dried.

The process of drying the fibers can take place in the sun or an oven under controlled temperature. This drying process is one of the important factor determining the mechanical performance characteristics of the extracted fiber. After drying, the fibers can be separated according to their length and fiber diameters. Good quality of fibers can be extracted by applying the mechanical decortication process. This process yields fibers of different lengths, for the same leaf that is subjected to decortication. For some fibers like flax after mechanical decortication additional steps such as hackling, which involves the separation of small and coarse fibers by passing (combing) the fibers through a comb-like structure known as hackle, are carried out. This step is followed by mechanical carding which untangles the fiber, cleans and mixes it to get it ready for future processing. Then, these steps are followed by the drawing and spinning processes. Sisal, Sansevieria cylindrica, bamboo, banana, hemp, PALF, etc., are some of the examples of the fibers extracted through the method of mechanical decortication.

3.1.3 Mechanical Dehulling

Husk/Hull based fiber products are generally extracted using this method. The process involves cursing or rolling of the seed or grains to separate the hull. This hull which is considered to be a waste by-product can be used as a reinforcement in composite materials. Green coconut coir fiber, rice husk, wheat husk, etc., are some of the examples of natural fibers or filler materials obtained through the process of mechanical dehulling.

3.2 Fiber Treatments

Fiber–matrix interface is one of the very crucial factors that determine the mechanical performance of the composite. This is especially true in the NFC's where most of the composite failures occur due to poor adhesion between the fiber surface and the matrix material. The reason behind this is that the load should be effectively transferred from the matrix material to the fiber material which is done through the fiber–matrix interface. The presence of excess contents of same constituents like wax would decrease the strength of the fiber. For the purpose of improving the physical and mechanical characteristics of natural fibers, different treatments can be utilized. There is a number of fiber treatment processes that can be used based on the improvement requirement and application. Surface treatment methods can broadly be classified as physical, physicochemical, and chemical treatments.

Physical treatments: Physical treatments are meant to modify the structural and surface properties of the lignocellulosic fibers. Fiber stretching, calendaring, rolling or swaging, solvent extraction, electric discharge, gamma-ray projection, and thermal treatments are the major physical modification methods.

Chemical treatments: Chemical treatments involve treating or subjecting the fibers to different chemicals for modifying the fiber surface and to remove some unwanted constituents like lignin and pectin. Alkaline, bleaching, peroxide treating, using coupling agents, enzymes are the main processes in the chemical treatments.

Physicochemical treatments: Physical treatments (thermal or steam explosion) combined with chemical treatments to enhance the chemical reactions that provide better fiber bundles are known as physicochemical treatments. These methods provide fiber or fibrils with very high cellulose content and also clean and fine fibers, whose mechanical properties are much closer to that of pure cellulose fibers, this delivers required improvements in the final composite material.

3.3 Composite Processing Methods

Manufacturing of biodegradable products forges to be the new trend in the development of composites. Hybrid biodegradable composite manufacturing pertaining to natural fiber composites is very much similar to the processing methods used to produce the conventional composites. Methods such as hand lay-up, resin transfer molding (RTM), vacuum molding, compression molding, injection molding, pultrusion, filament winding, etc. Out of the different methods that are used in the preparation of different types of hybrid composites, the hand lay-up and compression molding techniques are widely used for hybrid composites with thermoset matrices and injection molding and screw extrusion is the widely used methods for hybrid composites containing a thermoplastic matrix.

In composite materials, good surface refers to better mechanical properties. If the material has micro-crack or poor fiber–matrix bonding, then the material is said to have poor mechanical properties. Behavior and response of a material like viscoelastic behavior, impregnation, etc., has a connection with the processing technique applied for the production of the material and operation of the method of fabrication. If a composite does possess a foreign material in it then it is said to have inferior mechanical properties which are not a desired outcome of the processing methods. A brief discussion about some of the processing methods used for the production of hybrid composites is mentioned in the following paragraphs.

3.3.1 Hand Lay-Up Technique

One of the foremost open molding processes in the manufacturing of polymer composites is hand lay-up. It requires high skill for producing products. In this process, at first, mold release agents are sprayed to smoothen the surface of the mold for ease of handling of the composite thus produced. Both thermosets and thermoplastics can be used in this process. Usage of the optimum amount of the raw material for the required density of composite is done by estimating the mixing ratio of fiber to matrix. Automotive components are customarily produced using the hand lay-up technique due to its efficacy and accuracy. Casting comes under hand lay-up method.

3.3.2 Compression Molding

Compression molding is used for both thermosets and thermoplastic type of polymer composite manufacturing. Cold and hot processes are two different types of molding in this process. As the term compression is used to refer that usage of the pressure takes a major role in the process and no temperature is applied in the cold method. But the hot method demands the requirement of the usage of both pressure and temperature for the molding process. Application of heat and heat transfer in the hot method is used to initiate curing in the production process. Compression molding is generally used in the automotive industry which demands high production volumes. In order to achieve good distribution of reinforcements inside polymers, equipments such as internal mixers and twin screw extruders can be used.

3.3.3 Resin Transfer Molding (RTM)

RTM is a much popular form of production due its high production capacity and cheaper costs for producing components. Due to this reason, many automobile and aircraft companies use this technique to fabricate composites. In this process, first the fibers should be divided into fine pieces and the resin is injected on the fibers on a stationary platform, and then oven dried in a mold. Post curing is also required for this process.

3.3.4 Vacuum Molding

Preparation of prepregs through the molding method can be divided into vacuum bagging method and the autoclave method. The major diversity between these two processes is in the method of curing the matrix material. The plies of the laminate are consolidated and requires compaction pressure, which is applied through vacuum in the vacuum bagging process. First, the required prepreg materials are placed on a horizontal mold, and then covered by a vacuum bag. This vacuum bag is properly sealed to the mold by the use of the sealer materials. Then, using a vacuum pump, the air is drawn out of the vacuum bag to create the necessary vacuum for the matrix to cure. Many types of polymers such as epoxies, phenolic resins, and polyimides can be used and processed through this method.

3.3.5 Injection Molding

It is the most widely used process in the manufacturing of plastics components. The same technique is applied in the manufacturing of polymer composites with either natural or synthetic fibers. In order to produce composites through this process, the fibers have to be either small or in the powder form. The polymers have to be in the form of small pellets. A screw extruder can be used for consistent mixing of the matrix granules and the short fiber pieces or powdered fiber particles. The mixed granules flow through a hopper into the heating chamber, where the granules are heated, and then injected into the mold to take the shape of the mold at very high pressure. This is mainly applicable for producing the high-volume parts which are mainly made up from thermoplastic polymers.

3.3.6 Pultrusion

Pultrusion process can be defined as a combined process of both pulling and extruding. The fibers are impregnated with the resin to form an impregnated fiber composite by passing them through the resin bath and then these fibers move through a series of carefully designed dies in order to achieve the final shape of the product. Different shapes such as circular, square, rectangular, I and H shapes, etc., can be manufactured using this method.

3.4 Hybrid Composite Processing

Many researchers have produced different types of hybrid composites using several different methods which have been discussed in the following paragraphs:

Injection molding has been used by many researchers for producing hybrid composites. Nayak et al., and Samal et al., produced a hybrid composite containing short fibers of bamboo and glass as a reinforcement inside polypropylene matrix. They employed an intermeshing counter-rotating twin screw extruder for mixing the short fibers and the matrix pellets. Arbelaiz et al., prepared the hybrid composite containing flax fiber bundle and glass fibers as reinforcement in Polypropylene matrix. Panthapulakkal and Sain produced a hybrid composite containing short hemp and glass fibers as a reinforcement inside polypropylene matrix. Jarukumjorn et al., prepared a hybrid composite constituting of sisal and glass fibers in polypropylene matrix. Mirbagheri et al., prepared the hybrid composite containing wood flour and kenaf fiber-based hybrid composite reinforced inside a polypropylene matrix material. Tajvidi prepared the wood flour-based hybrid composite with equal amount of kenaf fibers reinforced inside a PP matrix material. Idicula et al., prepared the short fiber banana/sisal-based hybrid composite reinforced inside the polyester matrix. Noorunnisa khanam et al., prepared the hybrid composite material containing silk/coir that is incorporated inside an unsaturated polyester-based resin matrix material. These are some of the examples of the hybrid composites manufactured using the injection molding process.

Compression molding is one of the widely used techniques for manufacturing hybrid composites. This particular procedure can be used for processing both thermoplastics and thermosets polymeric composites. Thwe and liao prepared a short bamboo and glass fiber-based hybrid composite with a polypropylene matrix. Nayak et al., and Samal et al., prepared a short banana fiber and glass fiber-based hybrid composite. Kalaprasad et al., produced a short sisal and glass fibers-based hybrid composite with polypropylene matrix. Haneefa et al., prepared a hybrid composite made from short banana and glass fibers reinforced inside with polystyrene matrix. Zhang et al., prepared a flax and glass fiber-based hybrid composite with phenolic resin as the matrix. Saidane et al., prepared a flax and glass fiber reinforced inside epoxy matrix. Boopalan et al., prepared the hybrid composite containing jute/banana fiber-based epoxy composite. Venkateshwaran et al., prepared the hybrid composite containing short fiber reinforced sisal/banana fibers reinforced inside an epoxy matrix. These are some of the examples of the hybrid composites manufactured using the compression molding process. Senthil Kumar et al., prepared the hybrid composites containing short banana/woven coconut coir sheath reinforced inside the polyester matrix material. Athijayamani et al., prepared the hybrid composite containing sisal/roselle fibers reinforced in the polyester matrix material. Adekunle et al., prepared the hybrid composite containing flax fibers and lyocell fiber reinforced bio base thermoset aeso matrix. Paiva Junior et al., prepared the hybrid composite containing ramie/cotton fibers as the reinforcement inside the unsaturated polyester matrix material. Mwaikambo et al., prepared a hybrid composite containing cotton/kapok fibers reinforced inside an unsaturated polyester matrix material. De Medeiros et al., prepared the woven jute/cotton fiber-based hybrid composite material reinforced inside the novolac-type phenolic resin matrix material. Davoodi et al., produced a hybrid composite containing kenaf and glass fibers reinforced inside an epoxy matrix. These are some of the examples of the hybrid composites manufactured using the compression molding process.

Hot compression molding is another form of compression molding performed in assistance with the temperature. V. A. Patel et al., prepared a hybrid composite containing jute and carbon fiber-based hybrid composite with phenolic resin matrix. Mishra et al., prepared the hybrid composite containing PALF/glass fibers with polyester matrix. Amico et al., prepared a hybrid composite containing sisal/glass fibers reinforced inside the polyester. Zhong et al., prepared the hybrid composite containing sisal/aramid fibers in a phenolic resin matrix. These are some examples in which the researchers have used the technique of hot compression molding for fabricating the composites for their research.

Hand lay-up technique is one of the simple and most widely used techniques for producing composite materials. Santulli et al., prepared hybrid composites containing E-glass and flax fibers reinforced in epoxy matrix. Padma Priya and Rai had prepared a hybrid composite made from short silk and glass fibers with epoxy matrix. Ahmed et al., prepared a hybrid composite made up of jute (hessian cloth) and woven E-glass fiber mat with an unsaturated polyester matrix material. Ahmed et al., Ahmed and Vijayarangan, prepared a hybrid composite containing untreated jute woven fabric and glass fiber with an isothalic polyester matrix material. John et al., prepared a hybrid composite made up of sisal fiber and glass fiber reinforced in an unsaturated polyester matrix material. Khanam et al., prepared a hybrid composite made up of sisal fiber and carbon fiber reinforced in an unsaturated polyester matrix material. Srinivasan et al., prepared Banana/Flax fibers-based hybrid composite reinforced inside an epoxy matrix material. Raghu et al., prepared the hybrid composite containing sisal/silk fibers reinforced inside the unsaturated polyester (UPE) matrix. Alavudeen et al. prepared hybrid composites with pure woven banana and kenaf fibers reinforced inside the polyester matrix material. Kumar et al., prepared a hybrid composite consisting of the Jute/sansevieria cylindrica-based short fibers reinforced inside an epoxy matrix material. These are some examples in which the researchers have used the hand lay-up technique for fabricating the composites for conducting their research.

Hand lay-up followed by compression molding: This is a combination of both the hand lay-up and the compression molding, in which hand lay-up of the composite is done, and then followed by the compression modeling of the laminates or the composites. Sreekal et al., prepared the EFB/glass fibers-based hybrid composite with the phenol formaldehyde matrix. Shahzad et al., prepared a Hemp/glass fiber-based hybrid composites with polyester matrix. Khanam et al., prepared a hybrid composite with randomly oriented sisal and carbon fibers inside the unsaturated polyester matrix. Idicula et al., prepared a hybrid composite with randomly oriented short banana and sisal fibers inside the polyester matrix. Sarasini et al., prepared the hybrid composite containing basalt and carbon fibers reinforced in an epoxy matrix. Braga et al., prepared the hybrid composite containing carbon/basalt fibers reinforced in an epoxy matrix. These are some examples in which the researchers have used the hand lay-up followed by the compression modeling technique for fabricating the composites for conducting their research.

Casting method is one of the open molding technique, which can be used for effectively used for composite preparation. Venkata subba reddy et al., prepared the hybrid composites made up of bamboo and glass fibers reinforced inside a polyester matrix. Potham et al., prepared the hybrid composites made up of banana and glass fibers reinforced inside a polyester matrix. Venkat reddy et al., prepared the hybrid composites made up of bamboo and glass fibers reinforced inside a polyester matrix. Velumurugan et al., prepared the hybrid composites made out of palmyra and glass fibers reinforced inside a rooflite matrix. Raghavendra rao et al., prepared the hybrid composites made out of bamboo and glass fibers reinforced inside an epoxy matrix. Ashoke kumar et al., prepared the hybrid composites made out of sisal and glass fibers reinforced inside an epoxy matrix. These are some of the examples of the hybrid composites manufactured using the casting process.

Vacuum bagging technique is one of the important manufacturing technique used for preparation of composite performed in the presence of vacuum. Fiore et al. prepared the hybrid composites made up of Flax/basalt fibers reinforced inside the epoxy matrix. Ramesh et al., studied the hybrid composite material containing sisal/ jute fiber reinforced inside the epoxy matrix composite. Fiore et al., prepared a hybrid composite containing the carbon/flax with epoxy matrix vacuum bagging. These some of the examples of the hybrid composites manufactured using the vacuum bagging technique.

Resin Transfer Molding (RTM) method is another widely used method for producing structural laminates. Abdul Khalil et al., prepared the hybrid composite with EFB and glass fibers inside the polyester matrix and vinyl ester matrix. Resin impregnation method was used for producing the OPEFB/Glass fibers reinforced inside epoxy matrix by Hariharan et al., Idecula et al. prepared a PALF and glass fiber composites with polyester resin matrix. Jawaid M et al., prepared an OPEFB/ Glass fiber-based hybrid composites with epoxy matrix. These are some of the examples of the hybrid composites manufactured using the RTM technique.

Jute and glass fiber composites with unsaturated polyester were prepared using the pultrusion method by Zamri et al.

Relatively new technologies are the automated tape laying/automated fiber placement machines and creating the composites using the technology of 3D printing. Using the automated tape laying machine precise arrangement of prepregs can be performed on an open mold through the orientation and layup sequences, which were difficult to achieve before it became a reality. 3D printing of NFC's is being tested and the research is at a very early stage, but this particular method has a lot of potential implication on the field of composite material.

4 Properties and Characterization

4.1 Physical Properties

The physical properties of the natural fibers include the measurement of fiber diameter and density of the fibers. The dried fibers will have to be chopped into various lengths in order for us to measure the physical properties. Fresnel diffraction method can be used to generate the image of the fiber surface through the different diffraction patterns obtained from the surface of the fiber. Then, using the image processing softwares such as ImageJ, the diameter of the fibers can be determined. Bamboo, date, palm, banana, coconut, sisal, vakka, etc., fiber's diameters have been found out through this approach. Different natural fibers such as pineapple leaf fiber, coconut husk, kenaf, Sansevieria trifasciata, sisal, as abaca, ramie, etc., fiber diameters were tested using the optical microscope at a magnification scale of about 300×. The optical measurement was obtained at five different locations on the same fiber. Scanning electron microscopy (SEM) images were also utilized for the purpose of measuring the diameter of the various natural fibers with a magnification scale of up to 1000×. Comparisons were made between the measurements taken from both the microscopy method and the SEM image analysis method to find out the errors in measurements. In the SEM image method, image processing software's such as the ImageJ was used for outlining the cross section and shape of the fiber. Fibers such as sansevieria cylindrica, seagrass, hemp, okra, curaua, Phormium tenax, coir can be tested using SEM images. Profile projector and micrometer caliper equipment also used by some researchers to find out the diameter of the natural fibers such as curaua. Generally, a lot of readings will be taken and with the help of Weibull statistical analysis, the mean diameter of the fiber and standard error in measurement were identified. The density of the fibers such as sansevieria cylindrica, bamboo, banana, date, coconut husk, sisal, vakka, etc., was estimated through the pycnometric procedure. For some natural fibers such as hemp, kenaf, Sansevieria trifasciata, ramie, coconut husk, etc., the density of the fibers was estimated by weight to volume ratio method. Majority of the times, the pycnometric procedure will be applied to find out the density of the composite material.

4.2 Chemical Composition

Generally, the natural fibers are usually made up of cellulose, lignin, ash, wax, hemicellulose, pectin, moisture content, etc. In order to test the moisture content in the fiber, the difference in the weight of the raw and the dried fiber are compared. The difference in the normal weight between the normal fibers and the burnt fibers will result in identifying the percentage of ash present in the fiber. By subjecting the natural fiber to enzymatic degradation process and analyzing the galacturonic acid, the pectin content can be identified. Cellulose content in the fibers can be determined by the Kurshner and Hoffers method. Cellulose contents for the fibers such as kenaf, reed, miscanthus, switchgrass, cotton, etc., were identified using the above mentioned technique. X-ray (XRD) diffraction analysis can be used to find out the crystallinity index and values for the fibers, through which the cellulose content can also be determined. The lignin content can be determined for the natural fibers using the Klason lignin method. Lignin percentage of different fiber like sisal, seagrass, reed, miscanthus, cotton, and wood-based fibers were determined using the above mentioned method. The hemicellulose content of the fibers can be identified according to the NFT 12-008 method, in which the fiber will be heated in the presence of hydrobromic acid for extracting the hemicellulose. Wax content in a fiber can be determined using the Conrad method with Soxhlet extraction. Fourier transform infrared spectroscopy (FTIR) is one another method that can be applied to identify different contents in the fibers. On performing the FTIR analysis a spectrum with different bands will be generated and the analysis of the bands by comparing with standard available data, the contents of any material can be identified.

4.3 Mechanical, Thermal, and Dynamic Properties

Mechanical properties of the single fiber denote the tensile strength, tensile modulus, and percentage of elongation at breakage for an individual fiber. Coming to the composite materials, several properties such as the tensile, flexural, impact, compression, interlaminar shear strength (ILSS), wear, shear, fatigue properties are some of the mechanical properties that need to be tested. Different ASTM standards are specified in order to characterize the above mentioned properties for a fiber or a composite. The data such as dimensions of the test specimen, its arrangement in the fixture, required strain rates, precautions, limitations, and applicability are mentioned in the specific ASTM standards. For example, ASTM D 3379-75 standard specifies how to test and find out the tensile properties of the natural fibers. The thermal properties of the fibers or the composites can be understood with the help of the Thermogravimetric (TGA)/Differential Thermogravimetric analysis (DTG). The rate of thermal degradation can be found out using the TGA curves and the thermal decomposition temperatures can be identified using the DTG curves. Dynamic mechanical properties of the composites can be investigated by using the method known as Dynamic Mechanical Analysis (DMA). The characteristics such as the damping capability and the relative stiffness of a composite material can be identified through this method. The storage modulus, loss modulus, and damping factors for the composites can be obtained with respect to the time under the influence of several conditions such as different stresses, temperatures, frequencies, etc.

4.4   Properties of Hybrid Composite

Hybrid composites bring out most of the desirable properties of the combined materials. One reinforcement complements the disadvantages of the other reinforcements. When coming to the case of the hybrid composites prepared using the combination of natural and synthetic fibers, glass fiber is clearly a very preferable one, due its cost and properties, which can have good hybrid effects when combined with the low and medium strength natural fiber. A lot of researchers have done the same by characterizing the hybrid composites made out if different natural fibers with glass fibers.

4.4.1   Natural/Synthetic Fibers-Based Hybrid Composites

Sisal/Glass: Amico et al., studied the mechanical properties of pure glass, pure sisal, and hybrid sisal/glass composite with polyester matrix. It was concluded that the hybridization of the sisal/glass has produced intermediate properties between the pure glass and pure sisal properties. Jarukumjorn et al., studied the hybrid sisal/glass composites with PP matrix. The tensile, flexural, and impact strengths of the composite were increased and no significant change in the tensile and flexural modulus was observed with the advent of hybridization. Hybridization of sisal composite with glass fiber leads to improvement in the thermal and water resistance of the composite. John et al., studied the chemical resistance of the hybrid sisal/glass fiber with unsaturated polyethylene and found that hybridization has improved the chemical resistance of the composite except for the carbon tetrachloride. Misra et al., studied the mechanical properties of the sisal/glass fiber hybrid composites with polyester matrix. It was concluded that the introduction of small amount of glass fiber could potentially increase the tensile, flexural, impact, and resistance to water absorption properties of the hybrid composite in comparison with the unhybridized composite. Kalaprasad et al., evaluated the thermal conductivity and diffusivity of the hybrid short sisal/glass fiber with polyethylene matrix at various temperatures. The thermal conductivity of the composite increased with hybridization and vice versa for the thermal diffusivity. The thermal conductivity increased and leveled off with the increase in the temperature and the thermal diffusivity decreased with the increase in the temperature.

Sisal/Carbon: Khanam et al., studied the sisal/carbon hybrid composites reinforced inside polyester matrix. The tensile, flexural properties, and chemical resistance of the hybrid composite improved significantly in comparison with the unhybridized composites. The alkali treatment of the sisal fiber led to more pronounced improvements in the mechanical and chemical resistance properties of the hybrid composite. The hybrid composites were chemically resistant to all the chemical expect for CCL4.

Sisal/Aramid: Zhong et al., studied the hybrid composite containing sisal/Carbon fiber reinforced in a phenolic resin matrix. The effect of surface microfibrillation of the sisal fiber on the hybrid composite material was studied. The results showed a clear influence on the mechanical behavior of the hybrid composite materials due to the microfibrillation of sisal fibers. Due to the surface microfibrillation of the sisal fibers, the contact area between the sisal fibers and the phenolic matrix has increased drastically which in turn improved the mechanical properties. The tensile, compressive, matrix/fiber interfacial bonding strength and wear resistance of the hybrid composite have drastically improved.

Bamboo/Glass: Nayak et al. and Samal et al. studied the effect of incorporation of short bamboo fibers along with short glass fibers in a polypropylene (PP) matrix and indicated that there was an improvement in the stiffness and thermal stability of the composite. Thwe and Liao indicated that with the incorporation of up to 20 wt% of short glass (E-glass) fibers into short bamboo fiber reinforced polymer composite (BFRP) with PP as matrix material will increase its modulus and strength values in both the tension and bending in comparison with the normal BFRP composite. The durability of the BFRP composite with PP matrix has increased with the inclusion of short glass fiber (E-glass). They also have found that inclusion of short glass (E-glass) fibers increased the hygrothermal resistance and fatigue performance of BFRP composite at all the tested load levels. Raghavendra Rao et al. studied the effect of bamboo/glass fibers hybridization with the epoxy matrix on the flexural and compressive properties. It was concluded that both the modulus and strength values of the hybrid composite were more in comparison with the individual composites of either pure glass reinforcement or bamboo reinforcement, indicating a positive hybridization effect. Mandal et al., concluded that bamboo fibers can replace the glass fibers up to 25% without causing any significant change in the flexural and interlaminar shear properties of the hybrid composite prepared with unsaturated polyester and vinyl ester matrix. Venkata Subba Reddy et al., showed that the bamboo/glass fiber hybrid composite prepared using the polyester matrix has shown improved tensile properties and chemical resistance.

Kenaf/Glass: Davoodi et al., studied the effect of hybridizing the kenaf fiber with glass fiber with an epoxy matrix. It was concluded that the mechanical properties such as tensile strength, flexural strength, and modulus values of the hybrid composite have significantly improved. But the impact strength has decreased. It was concluded that positive hybrid effects were present and these composites can be used for structural components in cars. Atiqah et al., studied the treated kenaf/glass hybrid composite with UPE matrix and its applicability to structural usage. It was concluded that the hybrid composite containing 15% of each of the fiber will be yielding the highest mechanical properties.

Flax/Glass: Saidane et al., studied the jute/glass fiber hybrid composite with epoxy matrix. It was concluded that tensile modulus has increased and the tensile strength and specific tensile strength were decreasing due to hybridization. Santulli et al., studied the effect of inducing jute fibers as a replacement to E-glass fiber with the epoxy matrix on the impact properties of the hybrid laminates. Sufficient impact performance was observed along with weight savings with the hybrid laminate. Zhang et al. studied the mechanical properties of the unidirectional flax glass fiber hybrid composite with the phenolic resin matrix. The tensile modulus and strengths were improved with the hybridization. The fracture toughness and the interlaminar shear strength (ILSS) of the hybrid composite were high in comparison with the glass fiber reinforced composite. No hybridization effects were clearly observed. Arbelai et al., studied and compared different fiber treatments and matrix modification effects on the flax fiber/PP composite and the flax/glass/PP composites. Different chemical treatments using the chemical such as vinyltrimeethoxy silane, maleic anhydride–PP copolymer and fiber alkalization were carried out for the fiber and matrix modifications. It was concluded that the matrix modification yielded better mechanical performance rather than fiber treatment in both the flax/PP and the hybrid flax/glass/PP composites.

Flax/Carbon: Flynn et al., studied the effect of hybridization of the flax/carbon fibers-based hybrid composite with epoxy matrix. The hybridization of the flax composite has increased the mechanical properties of the composite material. The vibration damping of the composite has seen a decrease due to the hybridization. Fiore et al., studied the flax/carbon fiber-based hybrid composite with epoxy matrix mainly focusing on its mechanical behavior. The mechanical behavior was tremendously improved with an addition of carbon fiber lamina. The tests showed that this particular combination can be applied for various structural components in naval and automobile industries.

Oil palm empty fruit bunch (OPEFB)/Glass: Sreekala et al., studied the hybrid oil palm/glass composite with phenol formaldehyde matrix. With the inclusion of small amount of glass fiber, there was a significant improvement in the tensile and flexural properties of the composite. The presence of the OPEFB fiber improved the impact energy of the hybrid composite. Abdul Khalil et al., studied the mechanical properties of hybrid OPEFB/glass fiber composites with polyester matrix. It was concluded that hybridization leads to improvement in the tensile and flexural strengths of the composite in comparison with the pure OPEFB composite. Abdul Khalil et al., studied the mechanical properties of the hybrid oil palm/glass fiber composites with vinyl ester matrix. It was concluded that the hybrid composite has comparable properties with those of the pure glass fiber composite. Hariharan et al., studied the impact and tensile properties of the oil palm/glass fiber-based bilayer hybrid composite with epoxy matrix. The impact and the tensile strength were improved due to this hybridization.

Banana/Glass: Pothan et al., studied the effect of hybridization of banana fiber reinforced in a polyester matrix with glass fiber. An improvement in the tensile properties was observed with the addition of the glass fiber. It was also observed that as the percentage of glass fiber increased the tensile properties of the composite also have improved. There was an improvement that was observed in the impact strength properties of the hybrid composite when glass fiber was added up to an 11% and beyond that weight percentage, the impact strength of the hybrid composite decreased slightly. Haneefa et al., studied the short banana/glass fiber hybrid composites with polystyrene matrix. It was concluded that the tensile, flexural modulus and strengths of the composite were improving with the increase in the glass fiber percentage. But as the glass fiber percentage increased, there was a decrement in the percentage of elongation at breakage. Nayak et al. and Samal et al., studied the properties of short banana/glass fiber hybrid composite with PP matrix. The mechanical, dynamic, and thermal properties of the hybrid composites were improving with the hybridization keeping the ratio of banana to glass fibers at 1:1 at 30 wt% of both the combined fibers. The water absorption behavior of the hybrid composite has reduced significantly.

Jute/Glass: Ahmed et al., studied the effect of hybridization on the mechanical characteristics of jute and glass fiber (woven) with an isothalic polyester matrix. Significant improvement in the mechanical properties of tensile, flexural, and interlaminar shear strength (ILSS) was found with the inclusion of the glass fiber. This hybrid composite also offered better resistance to water/moisture absorption. Abdullah et al., studied a jute fiber and E-glass mat fibers reinforced inside the unsaturated polyester matrix-based hybrid composite. The properties have improved with the addition of the glass fiber, in comparison with the normal jute fiber reinforced composite. The improvement of the properties was significant when the ratio of the jute to glass fiber was 1:3. They also concluded that UV radiation treatment could further enhance the properties of the composite. Ahmed and Vijayarangan showed that hybridization of the jute composite with glass fiber with a polyester matrix significantly improved the flexural, tensile, and ILSS strength of the composite. The major effect was found on the upper and lowermost plies in the composite. Zamri et al., concluded water absorption causes a decrease in the flexural and compressive properties of the jute/glass fiber-based hybrid composite. Braga and Magalhaes clouded that hybridization of jute fiber-based epoxy composite with glass fiber has shown significant improvement in the impact energy, tensile, flexural, and density with a reduction in the moisture absorption of the hybrid composite.

Hemp/Glass: Panthapulakkal et al., studied the effect of hybridizing short hemp fiber reinforced inside a polypropylene (PP) matrix with glass fibers on the mechanical, thermal, and water absorption behavior of the hybrid composite. There was a clear indication that the tensile strength, tensile modulus, and impact strength of the unhybridized pure hemp composite were improved with the introduction of the glass fibers. The hybridization improved the thermal stability of the composite which was proved using the thermogravimetric analysis (TGA). The water absorption behavior of the hybrid composite was decreased in comparison with that of the unhybridized composite. Shahad studied the influence of short hemp/glass fibers on the impact and fatigue properties of the composite material with polyester matrix. It was observed that the impact strength increased for the hybrid composite if the replacement of hemp fibers with the glass fibers was up to 11%. The fatigue strength has increased for the hybrid composite, but the fatigue sensitivity of the hybrid composite remained somewhat similar to that of the unhybridized hemp fiber composite.

PALF/Glass: Misra et al., studied the mechanical properties of the PALF/glass fiber composite with polyester matrix. With the addition of the glass fibers, there was a significant amount of improvement in the tensile, flexural, and impact properties of the hybrid composite. The resistance to water absorption also increases with the hybridization, proving the existence of a positive hybrid effect. Idicula et al., studied the thermophysical properties of the short PALF/glass fiber-based hybrid composite with polyester matrix. Hybridization has led to the improvement in the effective thermal conductivity and heat transportability of the composites. Chemical treatment of the PALF fiber leads to even more improvements in the thermal conductivity of the composite.

Silk/Glass: Padma Priya et al., investigated the properties of silk/glass hybrid composite with epoxy matrix. The hybridization of the silk fiber epoxy composite with glass fiber showed an improvement in the mechanical properties of the hybrid composite. The water absorption behavior of the composite has decreased with the increase in the glass fiber percentage.

Palmyra/Glass: Velmurugan et al., studied the random palmyra/glass fiber-based hybrid composite with rooflite matrix with different fiber lengths and contents. It was shown that the maximum tensile flexural and impact properties at a fiber content of 54% and fiber lengths of 30–40 mm.

Basalt/Carbon: Sarasini et al., studied the properties of the Hybrid composite containing basalt/carbon fiber reinforced inside epoxy matrix. Mechanical analysis of the laminates made up with this layup has good mechanical properties due to the hybridization. the impact properties of the composite material with this layup has improved drastically.

4.4.2 Natural/Natural Fibers-Based Hybrid Composites

Sisal/Jute: Gupta et al., studied the hybrid composite material containing sisal/jute fiber reinforced inside the epoxy matrix composite. They studied the dynamic mechanical properties and water absorption behavior of the composite material. They concluded that the hybrid composite displayed high storage and loss modulus values. The damping properties of the hybrid composite have decreased in comparison with that of the unhybridized composite material. The water absorption behavior of the hybrid composite was much lower in comparison with that of the unhybridized composite.

Sisal/Roselle: Athijayamani et al., worked on the hybrid composite containing sisal/roselle fibers reinforced in the polyester matrix material. The study focused on the effect of hybridization on the tensile, flexural impact of the hybrid composite in both the wet and dry conditions. Good mechanical properties were obtained from the hybrid composite material. But the mechanical properties of the composite decreased significantly when the hybrid composite material was subjected to the moisture. The results showed that the degradation of the fiber and matrix interface was the main reason for the decrement in the mechanical properties of the hybrid composite when exposed to moisture conditions.

Sisal/Silk: Raghu et al., studied the hybrid composite containing sisal/silk fibers reinforced inside the unsaturated polyester matrix material. They showed that the hybrid composite with sisal/silk fibers was having considerable mechanical properties. The main improvement that occurred in the composite material with the introduction of hybridization is in the chemical resistance property. Chemical resistance tests were carried out on the hybrid composite material and they concluded that the hybrid material is resistant to different chemical except for the carbon tetrachloride chemical.

Banana/Sisal: Venkateshwaran et al., studied the hybrid composite containing short fiber reinforced sisal/banana fibers reinforced inside an epoxy matrix. The fibers were randomly oriented and the effect of hybridization on the mechanical and water absorption behavior was analyzed. The addition of sisal fiber to the banana fiber composite has shown an improvement in the mechanical properties of the pure banana fiber composite. The water absorbability of the hybrid composite material has decreased with the introduction of sisal fiber. All these results were up to a sisal fiber percentage of 50% by weight in the total fiber weight of the composite. Idicula et al., studied the effect of hybridization on the short fiber banana/sisal-based hybrid composite reinforced inside polyester matrix material on the static and dynamic mechanical properties. The study also focused their efforts on the thermophysical properties of the hybrid composites. The fibers were short and randomly distributed inside the matrix material. It was observed that with the hybridization, the effective thermal conductivity has decreased. It was also observed that the thermal conductivity of the composite has seen to increase with the inclusion of fiber which was chemically treated before the composite was manufactured. Higher flexural and impact properties were observed in the bilayer composite and the tensile strength is greater in the tri-layer composite with banana fiber layer as the top skin.

Banana/Kenaf: Alavudeen et al., studied the properties of the hybrid composites with pure woven banana and kenaf fibers reinforced inside the polyester matrix material. Positive hybrid effects were detected with this hybridization. The tensile strength, flexural strength, and the impact strength of the hybrid composite have shown a higher value in comparison with those composites with either pure kenaf fiber or pure banana fiber reinforcements.

Banana/Coir: Senthil Kumar et al. studied the hybrid composites containing short banana/woven coconut coir sheath reinforced inside the polyester matrix material. Hybridization of the composite with the higher percentage of the coconut sheath yielded significant improvements in the mechanical properties of the composite material. When the fibers are subjected to alkali treatments, a good effect on the mechanical behavior of the hybrid composite material was observed.

Banana/Jute: Boopalan et al., studied the hybrid composite containing jute/banana fiber-based epoxy composite. The influence of hybridization on the mechanical and thermal properties of the hybrid composite material was studied. The addition of banana fiber to the jute fiber composite increase the tensile, flexural and impact strengths of the hybrid composite. The thermal properties of the hybrid composite have also increased in comparison with the non-hybrid composite. The water absorption behavior of the composite has decreased.

Banana/Flax: Srinivasan et al., studied the mechanical and thermal properties of the banana/Flax fibers-based hybrid composite reinforced inside an epoxy matrix material. Positive hybridization effects were observed as the flexural strength of the hybrid composite has increased in comparison with the pure flax fiber composite or pure banana fiber composite. The similar trend was caught to be occurring in terms of the thermal properties of the composites. But, it is very difficult to accurately determine the thermal properties of the natural fiber composites in comparison with those of the synthetic fiber composites.

Cotton/Jute: De Medeiros et al., studied the woven jute/cotton fiber-based hybrid composite material reinforced inside the novolac-type phenolic resin matrix material. They found that the properties of the hybrid composites were largely dependent on multiple factors such as fiber content, fiber orientation, fiber–matrix interface, and fiber characteristics. Jute fibers were the major influencing material in composite materials. But the addition of the cotton to the jute will increase the breaking strength of the hybrid composite. Hybridization of jute with cotton fiber also helps to avoid a sudden catastrophic failure of the composite materials.

Cotton/Kapok: Mwaikambo et al., prepared a hybrid composite containing cotton/kapok fibers reinforced inside an unsaturated polyester matrix material. The effect of treating fibers with 5% NaOH solution was also studied. The effect of weathering on the mechanical properties of the hybrid composite was also studied. The increase in the volume fraction of the fiber leads to the cause in the decrement of the impact properties in the case of both treated and untreated fibers. The tensile strength of the hybrid composite with untreated fibers was high in comparison to that of the treated fiber-based hybrid composite. The tensile modulus of the alkali-treated fiber-based hybrid composite was higher in comparison with the hybrid composite containing untreated fibers. The flexural modulus and the strength values of the composite have decreased in when the hybrid fiber composites have been subjected to accelerated weathering conditions.

Cotton/Ramie: Paiva Junior et al., studied the hybrid composite containing ramie/cotton fibers as the reinforcement inside the unsaturated polyester matrix material. It was found from the results that there is no significant change in the properties of the ramie composite with the introduction of the cotton fiber. Poor alignment of the cotton fibers along with the poor fiber–matrix interface was given as the obvious reasons for the cotton fibers having no effect or negligible effect on the properties of the composite material. But these results, on the other hand, indicated that the ramie fibers are one of the good fibers that has very high potential to act as a reinforcement in a composite material.

Oil Palm Empty Fruit Bunch (OPEFB)/Jute: Jawaid et al., studied the hybrid composite containing EFB/Woven jute fiber reinforced in epoxy matrix material. They studied a tri-layer hybrid composite. The tensile and the flexural properties of the hybrid composite material were found to be less in comparison with the properties of the pure woven jute fiber epoxy composite. In a later study, Jawaid et al., studied the effect of jute fiber loading on the tensile and dynamic mechanical properties of the hybrid composite constructed with bilayer OPEFB/jute fibers reinforced in epoxy matrix composite. The tensile properties of the hybrid composite were observed to be on the higher side in comparison with that of the pure OPEFB epoxy-based composite. The tensile properties of the hybrid composite materials were increasing with the increase in the jute fiber percentage. The storage modulus of the hybrid composite is also increasing with the increase in the jute fiber percentage. A shift in the damping factor toward the higher temperature region was also observed with the increase in the jute fiber percentage of the hybrid composite material. The maximum amount of the tensile behavior and the damping behavior in the hybrid composite was observed at a ratio of 1:4 in the EFB/Jute fibers. It was finally concluded that the hybridization of the OPEFB fiber with jute fiber will be a beneficial combination in increasing the mechanical and dynamic properties of the pure OPEFB fiber-based epoxy composite. The reason behind the increase in the properties of the hybrid composite is that the fiber and matrix interface between the jute fiber and the epoxy is better in comparison with that of the interface between the OPEFB fiber and epoxy matrix. Finally, the jute fiber is a much more stronger fiber in comparison with the OPEFB fiber.

Flax/Basalt: Fiore et al., studied the composites made up of flax/basalt fibers reinforced inside the epoxy matrix. The hybridization of the flax composite with basalt has led to the reduction in the storage modulus of the hybrid composite after a completing a timeframe of 15 days. The critical temperature of decomposition has increased for the hybrid composite in comparison with that of the pure composite.

Wood Flour/Kenaf: Mirbagheri et al., experimentally investigated the tensile properties of the hybrid composite containing wood flour and kenaf fiber-based hybrid composite reinforced inside a polypropylene matrix material. It was concluded that the properties of the wood flour PP composite were increasing with the introduction of the kenaf fiber. Tajvidi studied the effect of hybridization of wood flour-based hybrid composite with equal amount of kenaf fibers reinforced inside a PP matrix material. Even after hybridizing the wood flour and PP composite with the kenaf fibers, the properties of the hybrid fiber were very much similar to that of the pure kenaf fiber composites. They concluded that the wood flour has a very minimum amount of reinforcing effect on the hybrid composite in comparison with that of the kenaf fibers. This also concluded that the wood flour is very much useful as filler material to reduce the voids, rather than a good reinforcement material, as it shows very minimum effect on the mechanical properties of the hybrid composite.

Coir/Silk: Noorunnisa khanam et al., has studied the hybrid composite material contains silk/coir incorporated inside an unsaturated polyester-based resin matrix material. Different lengths of fibers were taken and incorporated in the matrix material. The coir fibers were treated with NaOH solution and the effect of the treatment was also studied. Significant improvement in the tensile, flexural, and compressive properties of the hybrid composite were observed, and worth the induction of the chemically treated coir fibers. This study also showed good mechanical properties can be obtained from this coir/silk hybrid composite.

Jute/Sansevieria cylindrica: Kumar et al., studied the hybrid composite made out from the Jute/*Sansevieria cylindrica* short fiber composites reinforced inside with the epoxy matrix. Different fiber lengths were reinforced and the hybrid composite was prepared. The tensile and flexural properties were optimum at a fiber length of 2 cm and the compressive strength was optimum at a fiber length of 3 cm. The result showed that properties of the alkali treated fiber-based hybrid composite were high in comparison with the hybrid composite with untreated fibers.

Flax/Lyocell: Adekunle et al., studied the hybrid composite containing flax fibers and lyocell fiber reinforced bio base thermoset composite material. The hybridization of the composite material was introduced by sandwiching the carded lyocell mates in between the layers of the flax fabrics. It was concluded that the flexural properties of the composites were dependent on the thickness of the outer ply and the weave architecture of the embedded fibers. The flexural properties of the composite increased with the increase in the thickness of the outer ply. The major impact of hybridization of the flax fiber composite with the lyocell fibers was evident in terms of the resistance offered by the composite toward the moisture absorption. The water absorption behavior of the composite material has decreased drastically in the hybrid composite material in comparison with the composite reinforced with pure flax fibers.

The natural fibers and different hybrid composites can be tested for their physical, chemical, mechanical, thermal and dynamic properties using different above mentioned techniques. SEM imaging can also be effectively used for studying the fractured surface of both the natural fibers and the hybrid composites. Different failure phenomena such as fiber pull out, matrix failure, fiber delamination, and fiber–matrix interface have been effectively studied using SEM imaging.

Many of the researchers have put their efforts into determining the hybridization effects on the mechanical properties of the hybrid composites. They majorly studied the change in the properties of the hybrid composite with respect to the percentage of change in each fiber and the overall combined fiber percentage. They also studied the effect of different types of treatments applied to the fiber on the mechanical properties of the hybrid composites. Few people have also concentrated their efforts on characterizing the thermal and dynamic mechanical properties of the hybrid composites. It was observed that most of the hybridized fibers had a decrease in the amount of water absorption, in comparison with the pure fiber composites. Hybridization also in most of the cases tends to have improved the mechanical properties. Fiber treatments had a clear effect on the properties of the hybrid composites, which may be mainly due to the fact the treatments tend to improve the surface of the fiber and make the fiber–matrix interface stronger.

5 Applications of Natural Fiber Composites/ Bio-Composites (NFC's)

The application of NFC's in different field such as automobiles, aerospace, construction, sporting, etc., is definitely on the rise in recent decades, but far more research has to be conducted for on the NFC's to improve their number of applications. The use of the hybrid natural fiber reinforced composite as an effective replacement for the synthetic composites is far with only a few applications in

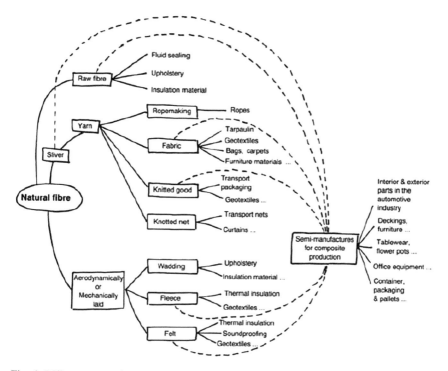

Fig. 6 Different conventional applications of the natural fibers in different forms [3, 4]

present. Figure 6 shows the different types of conventionally used for the natural fibers in different forms. Applications of pure and hybrid composites based on the natural fibers in different industries are discussed in the following paragraphs.

5.1 Automobile Industry

A tremendous rise in the usage of natural fiber composites in the last couple of decades have been observed, for use in different applications in different fields such as automobile components, sports equipment's, construction equipment, etc. European countries have put a lot of effort into increasing the usage of the natural fiber-based composites (both pure and hybrid NFC's) in the automobile components. Nonstructural components such as the door panels, instrumentation cluster, package trays, hat racks, boot liners, sun visors, internal engine covers, etc., and structural components where the load bearing is more crucial, such as exterior underfloor paneling, seat skeleton structure, etc., are being prepared with NFC's. The NFCs include hybrid composites such as wood/cotton fibers, flax/sisal fiber reinforced inside thermoset resins such as epoxy, polyester, etc., coconut fibers in either PP or PE matrix, leather/wool for interior textiles, in recent times America also has stepped into this particular area of using NFC's for different automobile

Fig. 7 Interior parts of the automobile of Mercedes Benz company made out of NFC's [3, 4]

components and is encouraging main international automobile players to adopt this strategy. The major players in the automobile industry have also increased their research efforts in making use of these NFC's for other structural and nonstructural composites in the automobiles. A major effort is being put on the use of natural fibers like hemp, jute, flax, sisal, and abaca fibers hybridized with either carbon or glass fibers for automobiles. Parts like indicator covers, L side covers, mirror coverings, etc., have been prepared from sisal and roselle fiber-based hybrid composite. Kenaf and glass fiber-based composites, Banana and glass fiber-based hybrid composites are being used as the material for constructing bumpers in modern cars. Banana fiber reinforced composites have been used for producing the mirror casings. Henry Ford has created full-body automobile made out of hemp and thermoset resins. Trabant was a car, whose body was constructed completely out of cotton fiber reinforced inside the polyester resin. Kenaf/hemp/wood-based hybrid composites used for door panels and inserts. Sisal fiber reinforced inside the epoxy matrix is used for creating cylinders for various applications. Figure 7 shows some automobile components that are used in the manufacture of benz company cars.

5.2   Aircraft Industry

In the initial days, when the aircraft industry was being developed, the first type of planes such as Havilland mosquito and Havilland albatross were manufactured from the composite plies made up of natural composites of balsa wood, plywood or untwisted flax fiber composites. But with the recent advent of the new class of hybrid composites known as the fiber–metal laminates (FML's) that are being used for the main structural components of the aircraft, due to their inherent advantage of possessing characteristics of both metal and fiber reinforced composite materials. The most common type of FML's used in the industry are aramid reinforced aluminum laminate (ARAL), carbon reinforced aluminum laminate (CARAL), glass fiber reinforced aluminum laminate (GLARE) which are mainly based on the combination of aluminum plates with different synthetic fiber reinforced composite plies. GLARE material is actively used as fuselage skin material in different aircrafts such as Airbus A380 and in impact resistant structures such as cargo area floor panels in Boeing 777 plane. ARAL is used as lower skin panels and cargo bay doors in airplanes such as Boeing C-17. CARAL with its high impact resistance is being applied in for helicopter structs and aircraft seat structures. The research is on the rise for replacing these synthetic fiber-based hybrid composite with NFC's. The aerospace industry has adopted the usage of the NFC's made from flax, hemp fibers for developing interior paneling for different aircraft. The hemp-based composite was studied for the application in developing electronic racks in the helicopters with very good results. Flax and glass-based hybrid composites with phenolic and other thermosets resins are being tested for use in small drones and fuselage skins for small aircraft and research on another such type of biohybrid composites are being undertaken by several aerospace agencies and companies such as Boeing, Airbus, etc.

5.3 Construction Industry

The field of construction has gone through several developments and familiarized itself for the usage of composite materials. Houses were constructed using the natural short straw fiber mixed with clay as the brick material from ancient time, which can be represented as a good example for natural fiber reinforced composite. The difficult nature of repair of the concrete structures and the challenging tasks associated with the restoration and replacement of old and heavy concrete or cement structures had encouraged the construction industry to adopted the strategies of using carbon fiber reinforced polymer (CFRP) and glass fiber reinforced polymer composites (GFRP). Despite the advantages in the synthetic fiber-based polymer composites, recycling of these materials is not possible. The construction industry has been testing the applicability of basalt, hemp fibers as a second phase of reinforcement inside concrete in addition to the steel frame. Straw-based construction materials are also being developed and used in the USA for the purpose of construction. Bamboo is a very widely used construction material, especially in the Asian countries. New bamboo and glass fiber-based hybrid composites were developed and are being used for producing artificial panels with high strength and termite resistance. Composites made from soy oil-based resins and cellulose fibers are being used for the construction of structures. The chopped sisal/jute fibers

Fig. 8 Window frames and paneling made from wood, flax, rice husk bagasse fibers-based hybrid composites [3, 4]

reinforced inside polyester resins are being used to develop laminates used for interior paneling inside buildings and offices.

Door frames are being prepared with the jute hessian clothe reinforced in phenolic resins produced through pultrusion technique. wood is one other majorly used hybrid composite in the field of construction and good hybrid composites from different wood and wood/glass fibers are being developed with very high impact and flexural strength to be used as a structural material for construction of temporary constructions, which can easily be disassembled and moved to where it has to again assemble for temporary usage. New composite materials made up of wood, flax, bagasse, rice husk, and their hybrids are used in the development of insulation panels, floor panels, fencing bars in the construction. Coir based polyester composite materials are used for producing post boxes, paperweights, roofing panel solutions, etc. Sisal and glass fiber base hybrid composites which are manufactured through the process of high-pressure compression molding are used as replacement for asbestos—cement sheets (Fig. 8).

5.4 Sporting Industry

Natural fiber composites are being used in the sporting industry for producing sports equipments such as bicycle frames, rackets, fishing poles, snowboards, golf rods, etc. Museeuw bikes are one of the first, race bikes which had uses the flax and carbon fiber-based hybrid composite with reinforcement inside the epoxy matrix for the development of bicycle frames. They also have developed the hemp and flax-based hybrid composites for the bicycle frames and other parts for the racing bikes. Snowboards are being prepared from the hemp fiber-based NFC's. Hemp and glass fiber hybrid composites are also being evaluated for different sporting equipment. Flax, carbon hybrid composites are also used for the development of golf clubs and fishing rods. Coir fiber-based composite materials are used for producing sporting and other types of helmets. "Ecoboard" was developed with the hemp fiber and bio-based resin and is being currently used as the material for manufacturing surfing boards. Fishing rods were developed by a company named "Cellu Comp", using the cellulose that has been extracted from the beetroot plants.

5.5 Electronics Industry

Kenaf fiber reinforced inside polylactic acid (PLA) matrix has been used to produce the outer casing of the mobile phones. "N701iECO" is the first model of eco-friendly devices developed by NEC Corporation. Flax fibers-based natural fiber composites are being used to replace the plastic chassis for the laptops. Dell has used the bamboo fiber based natural composite materials for replacing the aluminum body of its premium laptop series. Hemp and glass fiber reinforced hybrid composite materials are being used as the rack material for hosting servers and other computer equipments. Banana fiber reinforced composites have been used for producing the projector casings. Banana and epoxy-based composites are being used in the electrical industries. A lot of research on how to use natural fiber reinforced composite polymeric materials for different parts of electronics and electronic packaging is being undertaken.

5.6 Transport

In India, wood, bamboo, and glass fiber-based boards are being developed as an alternative material to replace the medium density fiber board panel which is being tested for the use in the body of the rail cars. In Germany, green buses made out composite from natural fibers like flax, wood, hemp, sisal, coir are being used for development of hybrid composites for the use in the bodies of trucks and buses. Different parts for trucks, buses, and locomotives are being researched for effective utilization of hybrid natural fiber-based composites.

5.7 Energy Sector

Flax fiber reinforced polyester composite was developed and manufactured using the resin transfer molding process (RTM) by JAC composites group, which was used to produce a high-performance turbine blade (Fig. 9). Studies are being undertaken on the use of the bamboo fiber for producing the turbine blades. Generally, glass fiber reinforced polymer (GFRP) and carbon fiber reinforced polymer (CFRP) composites are used for the production of the wind turbine blades. But research has been going on the use of basalt fiber-based hybrid composites containing either glass fibers or carbon fibers for use in wind turbine blades are being conducted. Several natural fibers like hemp, flax, jute-based pure and hybrid composites are being evaluated for the use in the energy sectors.

Fig. 9 Wind turbine blade made out of flax fiber reinforced inside the polyester matrix [5]

Many other applications of natural fibers and their hybrid composites are being researched and developed for the use in many other applications. The overall global market for the natural fiber-based composites is steadily on the increase. This steady rate of growth in the research and development of natural fiber-based composites is an indication of the potential applications of these types of composites.

In the recent times, manufacturers are looking toward more sustainable alternatives in every aspect of their product design and manufacturing as customers are also increasingly showing interests for these sustainable products. Despite having a good number of desirable characteristics, the natural fiber reinforced composites present a number of challenges. The high amount of water absorption, low thermal properties, lower mechanical durability, lack of techniques to process and produce high quantities of fiber with homogeneous characteristics, some fibers needing excessive treatments with higher costs are some of the challenges in adapting natural fiber-based composite materials for huge scale industrial applications. In order to overcome some of these challenges hybridization of composite looked as one of the feasible solutions. Research has to be continued on the topic of increasing the thermal and water resistance of the natural fibers. Researchers should also concentrate on developing sustainable, standardized processing and production techniques for producing natural fibers with desirable and homogeneous properties. Lot of automobiles, aerospace, construction, and sports industries are majorly investing in the usage of natural materials. Exploitation of the bio-based unused materials as a filler material in these natural fiber reinforced composites is also being gradually researched upon. Attention toward the development of polymers from biowastes and starches obtained from various natural sources is also another area of concentration and researchers have been successful in producing some natural polymers. Research should be continued on the development of more sustainable polymers derived from the natural sources that can be used for natural fiber composites. Future of these particular fibers, polymers, and materials is looking very bright.

6 Conclusion

A lot of research efforts are being put forth on the development of natural fiber-based hybrid composites as an effective alternative to synthetic fiber composites. Many researchers have studied the hybrid composites made up of natural/synthetic and natural/natural fibers. Most of the researchers have concentrated their effort on the prediction the mechanical property variations with respect to the change in total fiber percentage and change of one fiber percentage in the hybrid composite. Glass fiber was found to be a good synthetic fiber that can be used for hybridizing the natural fiber composites.

Compression molding and hand lay-up are the most often used processing methods for producing hybrid composite with thermoset matrix and injection molding is the most often used processing method for the hybrid composite with thermoplastic matrix material. Most of the natural fiber composites are being widely applied in the construction field followed by the automobile fields. Lot of other fields are also conducting research in the area of NFC's for developing novel pure and hybrid composite to be used in the various applications.

References

1. Bassyouni M et al (2017) Bio-based hybrid polymer composites: a sustainable high-performance. In: Hybrid polymer composite materials processing. Woodhead Publications, pp 23–70
2. Saba N, Jawaid M (2017) Epoxy resin-based hybrid polymer composites. In: Hybrid polymer composite materials: properties & characterization. Woodhead Publications, pp 57–82
3. Nguyen et al (2017a) Mechanical properties of hybrid polymer composite. In: Hybrid polymer composite materials: properties & characterization. Woodhead Publications, pp 83–114
4. Nguyen H et al (2017b) Hybrid polymer composites for structural applications. In: Hybrid polymer composite materials: applications. Woodhead Publications, pp 35–52
5. Shah DU, Schubel PJ, Clifford MJ (2013) Composites: part B can flax replace E-glass in structural composites? A small wind turbine blade case study. Compos B 52:172–181
6. Biagiotti J, Puglia D, Kenny JM (2004) A review on natural fibre-based composites—part II. J Nat Fibers 1(2):37–68
7. Chang Hong R, Wood (2004) A review on natural fibre-based composites-part I: structure, processing and properties of vegetable fibres. J Nat Fibers 1(2):37–41
8. Dittenber DB, Gangarao HVS (2012) Critical review of recent publications on use of natural composites in infrastructure. Compos A Appl Sci Manuf 43(8):1419–1429
9. Dong C (2018) Review of natural fibre-reinforced hybrid composites. J Reinf Plast Compos 37(5):331–348
10. Fuqua MA, Huo S, Ulven CA (2012) Natural fiber reinforced composites. Polym Rev 52(3–4):259–320
11. Guna V et al (2017) Hybrid biocomposites. Polym Compos 1–25
12. Kan F, Zheng L, Potluri R (2016) Buckling analysis of a ring stiffened hybrid composite cylinder
13. Kanitkar YM, Kulkarni AP, Wangikar KS (2017) Characterization of glass hybrid composite: a review. Mater Today: Proc 4(9):9627–9630

14. Kistaiah N et al (2014) Mechanical characterization of hybrid composites: a review. J Reinf Plast Compos 33(14):1364–1372
15. Pickering KL, Efendy MGA, Le TM (2016) A review of recent developments in natural fibre composites and their mechanical performance. Compos A Appl Sci Manuf 83:98–112
16. Potluri R (2018) Mechanical properties evaluation of T800 carbon fiber reinforced hybrid composite embedded with silicon carbide microparticles: A micromechanical approach. Multidiscipline Model Mater Struct. https://doi.org/10.1108/MMMS-09-2017-0106
17. Potluri R, Diwakar V et al (2018) ScienceDirect analytical model application for prediction of mechanical properties of natural fiber reinforced composites. Mater Today: Proc 5(2):5809–5818
18. Potluri R, Eswara A et al (2017) ScienceDirect finite element analysis of cellular foam core sandwich structures. Mater Today: Proc 4(2):2501–2510
19. Potluri RKJP et al (2017) ScienceDirect mechanical properties characterization of okra fiber based green composites & hybrid laminates. Mater Today: Proc 4(2):2893–2902
20. Potluri R, Dheeraj RS, Vital GVVNG (2018) ScienceDirect effect of stacking sequence on the mechanical & thermal properties of hybrid laminates. Mater Today: Proc 5(2):5876–5885
21. Potluri R, Paul KJ, Babu BM (2018) ScienceDirect effect of silicon carbide particles embedment on the properties of Kevlar fiber reinforced polymer composites. Mater Today: Proc 5(2):6098–6108
22. Potluri R, Rao UK (2017) ScienceDirect determination of elastic properties of reverted hexagonal honeycomb core: FEM approach. Mater Today: Proc 4(8):8645–8653
23. Ramesh M, Palanikumar K, Reddy KH (2017) Plant fibre based bio-composites: sustainable and renewable green materials. Renew Sustain Energy Rev 79(May):558–584
24. Rowell RM, Han JS, Rowell JS (2000) Characterization and factors effecting fiber properties. Nat Polym Agrofibers Compos 115–134
25. Safri SNA et al (2018) Impact behaviour of hybrid composites for structural applications: a review. Compos B Eng 133:112–121
26. Sathishkumar T, Naveen J, Satheeshkumar S (2014) Hybrid fiber reinforced polymer composites—a review. J Reinf Plast Compos 33(5):454–471
27. Sathishkumar TP et al (2013) Characterization of natural fiber and composites—a review. J Reinf Plast Compos 32(19):1457–1476
28. Singh J et al (2017) Properties of glass-fiber hybrid composites: a review. Polym Plast Technol Eng 56(5):455–469
29. Thakur VK, Thakur MK, Gupta RK (2014) Review: raw natural fiber-based polymer composites. Int J Polym Anal Charact 19(3):256–271
30. Väisänen T, Das O, Tomppo L (2017) A review on new bio-based constituents for natural fiber-polymer composites. J Clean Prod 149:582–596
31. Zabihi O et al (2018) A technical review on epoxy-clay nanocomposites: structure, properties, and their applications in fiber reinforced composites. Compos Part B: Eng 135:1–24. Available at: https://doi.org/10.1016/j.compositesb.2017.09.066
32. Potluri Rakesh (2018) Mechanical properties of pineapple leaf fiber reinforced epoxy infused with silicon carbide micro particles. J Nat Fibers. https://doi.org/10.1080/15440478.2017.1410511
33. Asim M et. al (2017) Processing of hybrid polymer composites—a review. In: Hybrid polymer composite materials processing. Woodhead Publications, pp 1–12
34. Ashori A (2017) Hybrid thermoplastic composites using non-wood plant fibers. In: Hybrid polymer composite materials: properties & characterization. Woodhead Publications, pp 39–56
35. Nurul Fazita MR et al (2017) Hybrid bast fiber reinforced thermoset composites. In: Hybrid polymer composite materials: properties & characterization. Woodhead Publications, pp 203–234

Processing of Green Composites

M. Ramesh and C. Deepa

Abstract The research on environmentally friendly materials has been regressively carried out to protect the environment. Natural fiber reinforced green composites have created enthusiasm for an extensive variety of research exercises. Many researchers are attempting to create different products by utilizing natural fibers as reinforcing agents. It is the main reason that this chapter is outfitted through reviewing the published results. This chapter draws out the issues related with processing of natural fibers and their green composite has been tended to. The types of fibers utilized as a part of composites, surface modifications and processing techniques of these green composites have been examined in detail. From the literature, it is presumed that the natural fibers are the potential replacement option for unadulterated manufactured fibers and the utilizations of these fiber composites will increase in future.

Keywords Natural fibers · Energy efficient · Bio-degradability
Surface modifications · Processing methods

1 Introduction

Growing environmental awareness and increasing interest in sustainability have led to the development of bio-composites for structural applications [1, 2]. The viability of natural issues urged the need to search for new choices which could alternate the regular composites with bring down environmental impact [3–6]. These made an enthusiasm for common materials which could be utilized as reinforcing materials

M. Ramesh (✉)
Department of Mechanical Engineering, KIT—Kalaignarkarunanidhi Institute
of Technology, Coimbatore 641402, Tamil Nadu, India
e-mail: mramesh97@gmail.com

C. Deepa
Department of Computer Science and Engineering, KIT—Kalaignarkarunanidhi Institute
of Technology, Coimbatore 641402, Tamil Nadu, India

© Springer Nature Singapore Pte Ltd. 2019
S. S. Muthu (ed.), *Green Composites*, Textile Science and Clothing Technology,
https://doi.org/10.1007/978-981-13-1972-3_2

in composites and are along these lines called as green composites or eco-composites or bio-composites [7–9]. Natural fibers are renewable resources with many advantages. They are abundant, inexpensive, light in weight, strong and non-abrasive, they can serve as an excellent reinforcing agent for plastics to replace or to reduce utilization of synthetic fibers in different applications [3, 4, 6, 10, 11]. Thus, a significant number of works has been done in this regard in recent years [12–22]. Numerous experiments were conducted by reinforcing natural fibers into polymer matrix to create composites for applications which don't require incredible strength, for example, auxiliary building structures, auto entryway boards, bundling, and so forth [6].

Green composites are a type of bio-composites where both the matrix and/or reinforcement are based on cellulose [23]. In addition to their bio-degradability, the benefit of these composites is the chemical similarity of the matrix and the reinforcement, whereby they are able to overcome the problem of poor fiber-matrix adhesion which is common in bio-composites [24, 25]. Because of increasing ecological effect, natural fibers have been widely recognized as alternatives for synthetic fibers as reinforcements in composites for decades [26–31]. The utilization of natural fiber reinforced green composites in many applications is because of broad research attempted and excellent properties such as low density, great strength, handling adaptability, high stiffness, and so forth [4, 8, 32–40]. Reports demonstrated that natural fiber based composites have picked up in significance importance as the reinforcing component in composites [19, 41–43]. Therefore, this chapter focuses an overview of recent development on the use of bio-fibers and their composites, for engineering and other industrial applications. Firstly, the fibers, their extraction and separation processes are discussed. Next, the fiber surface modifications are presented. Then the manufacturing processes of green composites are discussed in detail. Finally, the chapter has been concluded with the findings of this literature.

2 Natural Fibers

Synthetic fibers have got more specific strength compared to natural fibers, they are not eco-friendly and economical, which has resulted in the researchers showing much interest in these fibers [44, 45]. Owing to their low weight-to-high strength ratio and recyclable features, the natural fibers are the most potential choice in place of synthetic fibers and been used as reinforcement materials in polymer matrix composites [46]. Faruk et al. [3] summarized the development of bio-composites and reported that the plants to be utilized as fibers for essential utilities while the byproducts of plants have a place with the auxiliary purposes. Table 1 demonstrates the amount of few natural fibers produced every year and their topographical dispersion [4, 6, 47–54]. The chemical composition of these fibers is cellulose, hemi-cellulose, lignin and the amount of these parts change from plant to plant. This is because of age, species, and could likewise change in various parts of a plant.

Table 1 Natural fiber production and their major producers

Fiber	World production ($\times 10^3$ tons)	Major producers
Abaca	70	Costa Rica, Ecuador, Philippines
Bagasse	75,000	Brazil, India, China
Bamboo	30,000	China, India, Indonesia
Coir	100	India, Sri Lanka
Cotton	25,000	China, India, USA
Curaua	0.9	Brazil, Venezuela
Flax	830	Belgium, Canada, France
Hemp	214	China, France, Philippines
Henequen	0.03	Mexico
Jute	2300	Bangladesh, China, India
Kenaf	970	Bangladesh, India, USA
Oil palm	40	Malaysia, Indonesia
Pineapple	74	Indonesia, Philippines, Thailand
Ramie	100	Brazil, China, India, Philippines
Silk	0.10	China, India
Sisal	378	Brazil, Tanzania

These essential segments halfway decide the properties of the fibers and its composites. Cotton has the most astounding measure of cellulose while coir fiber has the most astounding measure of lignin. The physical properties of various natural fibers have been reported in Table 2.

2.1 Fiber Extraction and Separation Process

The natural fibers utilized as a part of commercial applications, for the most of the fibers are isolated from the stem of the respective plant by the retting procedure. Retting is a procedure which isolates the fiber groups from central stem which extricates the fibers from woody tissue of the plants. The fiber separation affects fiber quantity, quality of fiber, chemical mixture, fiber structure, and its properties [76]. The classification of retting processes is presented in Fig. 1. It can be isolated into four main partition procedures, for example, biological, mechanical, physical, and chemical process. Biological process, for example, bacteria and fungi in the earth assumes a noteworthy part in the debasement of the pectic-polysaccharides from non-fiber tissue and isolated fiber groups. Here and there, the retting procedure could be trying concerning the alert required as under-retting brings about polluted fibers [76]. The typical extracted fibers are exhibited in Fig. 2 [38, 76–79].

Table 2 Physical properties of various natural fibers

Fiber	Density (g/cm^3)	Diameter (mm)	Refs.
Abaca	1.50	–	[58]
Abaca leaf fiber	0.83	114–130	[57]
Agave	1.20	126–344	[59]
Alfa	0.89	–	[58]
Bagasse	1.25	200–400	[58, 60]
Bamboo	0.91, 0.6–1.1	240–330	[58, 61]
Banana	1.35	60–80, 50–250	[61, 62]
Coir	1.15, 1.15	100–160, 100–450	[61, 62]
Coconut tree leaf sheath	–	140–990	[63]
Cocount husk fiber	0.87	117–125	[57]
Cotton	1.5–1.6	–	[68]
Curaua	1.40	170	[58, 64]
Date	0.99	155–250	[61]
Date palm	1.20	–	[58]
Elephant grass	0.817	70–400	[65]
Flax	1.4–1.5	–	[66, 68]
Hemp	1.48	–	[68]
Henequen	1.20	–	[58]
Isora	1.30	–	[58]
Jute	1.3–1.46	40–350	[60, 68]
Kapok	1.47	22–65	[67]
Kenaf	0.25	70–250	[55]
Kenaf bast fiber	1.31	65–71	[57]
Oil palm	1.55	–	[58]
Okra	–	61–93	[68]
Palm	1.03	400–490	[61]
Palmyrah	1.09	20–80, 70–1300	[56, 69]
Petiole bark	0.69	250–650	[56]
Phormium tenax	–	100–200	[70]
Piassava	1.40	–	[58]
Pineapple leaf fiber	1.32, 1.44	53–62, 20–80	[57, 69]
Rachilla	0.65	200–400	[56]
Rachis	0.61	350–408	[56]
Ramie	1.50	50	[48, 60]
Ramie bast fiber	1.38	46–54	[57]
Rice husk	0.50	70–250	[55]
Root	1.15	100–650	[56]
Sansevieria cylindrical	0.915	230–280	[71]
Sansevieria ehrenbergii	0.887	10–250	[72, 73]
Sansevieria leaf fiber	0.89	83–93	[57]

(continued)

Table 2 (continued)

Fiber	Density (g/cm^3)	Diameter (mm)	Refs.
Sea grass	1.50	5	[74]
Sisal	1.45	50–300	[48, 62]
Sisal leaf fiber	0.76	122–135	[57]
Softwood kraft	1.50	–	[68]
Spatha	0.69	150–400	[56]
Talipot	0.89	200–700	[56]
Vakka	0.81	175–230	[61]
Wheat straw	1.45	–	[75]

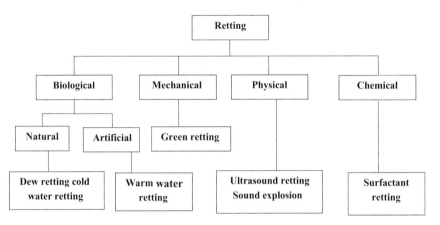

Fig. 1 Classification of retting processes

2.1.1 Dew Retting

This is also called field retting, is the most established and most generally utilized retting procedure to isolate fibers from the plants. This procedure requires proper moisture and surrounding conditions, and accordingly can't be utilized all around globally. The plant stays in the field in the wake of gathering for the miniaturized scale living beings to isolate fibers. The plants are turned over all the time to guarantee homogeneous retting. The retting procedure ought to be observed and halted at the ideal time to anticipate degradation of cellulose by smaller scale; and this is called as over-retting. The over-retting diminishes the mechanical execution of the fibers while under-retting makes further fiber preparing troublesome. Dew retting relies upon climate conditions and regularly takes 3–6 weeks. There are a few weaknesses in this procedure, for example, low quality of fibers contrasted with different procedures, dirty dull fibers because of contact with soil restricted to topographical area, wild climate conditions, and occupation of land until the retting is finished [76].

Fig. 2 Extracted natural fibers **a** hemp; **b** kenaf; **c** jute; **d** borassus fruit; **e** coir; **f** wood; **g** corn; **h** grass; **i** roselle; **j** sisal; **k** banana; **l** palmyra; **m** pineapple; **n** abaca; **o** cotton; **p** areca husk; **q** snake grass and **r** henequen

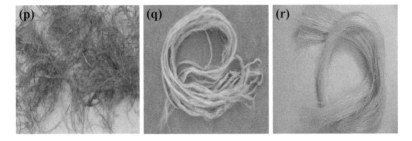

Fig. 2 (continued)

2.1.2 Stand Retting

An altered field retting was endeavored to overcome the restrictions of dew retting is called stand retting. A pre-reap desiccant, glyphosate was utilized to encourage retting. It was discovered that fungal spread and retting was slower than dew retting yet in comparison the fibers would be better quality when the stand retting strategy was utilized [80]. Thermally induced stand retting strategy is another stand retting. In this process, the plant development is ended by open gas flames and the plant bases are heated roughly 100 °C. The plants dry in two or three days which permits the substitution of weather-vulnerable dew retting. The danger of climate and crop damage is decreased by utilizing this strategy however the cost of retting increases [6].

2.1.3 Cold Water Retting

This retting includes soaking the fibers in running water, for example, waterways or streams. Afterward, this was replaced by water tanks which are fixed or opened relying on the weather and the water change happens every 2 days. This retting regularly takes 1–2 weeks, however the span generally relies upon the temperature of the water, the water quality, and the bacterial capacity. This retting procedure delivers high quality fibers than field retting. There are a few disadvantages of this procedure, for example, high quantity of water utilization, high cost of artificial drying in the wake of retting, natural contamination from waste water, and rankness of maturation gasses. This procedure has been ended in many parts of the world because of ecological concerns [76].

2.1.4 Warm Water Retting

This is a quickened water retting process which delivers clean and high quality of fibers within a week. The water in the reservoir is warmed around 40 °C. The procedure and disadvantages are like that of cold water retting, and has been canceled in many parts of the world. The waste water could be utilized as fluid manure in the event that it is dealt with to expel poisonous components [76].

2.1.5 Mechanical Retting

Mechanical retting, otherwise called green retting, is a simple and economic procedure to isolate the fibers from the plants. This retting is performed after field retting or specialized drying. Field drying ordinarily takes around 48–72 h. In this process the fibers are isolated from woody tissues mechanically. Fibers isolated by mechanical retting are coarse than field or water retting process [76].

2.1.6 Steam Retting

This method is a good contrasting option to the conventional retting process. The steam and the added substances enter the fiber interspaces of the fiber packages under pressure and high temperature; that expels the inside lamella at ideal conditions. The subsequent unwinding prompts separating of the fibers and results in decay into fibers. The fibers separated by this strategy are very soft and have great mechanical and physical properties [76].

2.1.7 Enzyme Retting

The pectin degrading chemicals are utilized to isolate the fibers from the woody tissue of the plant. This gives controlled retting of the fiber crops through specific bio-degradation of the pectinaceous substances. The enzyme action could increment with expanding temperature up to an ideal temperature above which the chemical begins to denature. The fibers separated by this procedure are of high and consistent quality [76].

2.1.8 Chemical and Surfactant Retting

Chemical retting refers to all retting forms in which the fibrous area of the plant is sub-merged in warmed water tanks with sulphuric acid, chlorinated lime, sodium or potassium hydroxide solutions and soda ash to break up the pectin part. The surface dynamic agents could be utilized as a part of retting to remove the undesirable non-cellulosic segments sticking to the fibers. The fibers produced by this process are top-notch quality yet the cost is higher than that of customary methods [76].

2.1.9 Osmotic Degumming

The Spanish broom and flax plants were placed in a glass gauge filled with warm water and placed in a tank filled with water heated about a temperature around 35 ° C. One end of the tube was immersed in the glass gauge and the other end was immersed in a container. This method of maceration uses natural physical laws such as water diffusion, osmosis and osmotic pressure. Osmotic degumming of the Spanish broom plant lasted 28 days and for flax 3 days, after which fibres were extracted by mechanical processes such as breaking and scotching [81].

2.2 Fiber Surface Modification

So as to acquire dependable composite materials for mechanical applications and to use completely the capability of reinforcing fibers, both perfect reinforcement and solid interfacial bonding arrangement must be ensured [82, 83]. The properties of natural fibers can be enhanced by suitable treatment techniques, especially considering woven fibers in fabric forms, for composite manufacture [84–88]. Notwithstanding, the use of natural fibers has a few disadvantages, for example, the hydrophobicity of the fibers, the moderately poor thermal dependability of the bio-composites and particularly the poor compatibility towards a hydrophobic matrix, bringing about weak interfaces and poor properties of the composites. The greater part of these drawbacks might be overwhelmed by the utilization of surface treatment of fibers [89–91]. Then again, being a natural material, the natural fibers are vulnerable against organic assaults, influenced by the moisture content and sensitive to alkaline solution which can decompose lignin and other constituents. Hence, a few solutions were proposed in the literature so as to counteract degradation phenomena in fibers [92–94]. The surface treated fiber reinforced composites are serves better as far as corrosive resistance, and other attractive properties when compared with the conventional materials [95]. The shortcomings of these composites can be enhanced by modifying the fibers by means of chemical or physical strategies or by the utilization of coupling agents [96–98].

2.2.1 Physical Treatments

The distinctive physical modifications like boiling of fiber with or without pressure can remove the impurities on fiber surface which can respond with resin effectively to form a solid interface [99]. The resin coatings on the fiber surface with phenol formaldehyde or resorcinol formaldehyde by various methodologies are pro-fondly successful in upgrading the reinforcing nature of fiber, giving as high as 40% enhancements in flexural property and around 60% changes in the modulus. These changes enhance the interfacial relationship between the fiber and matrix, resin wettability, and so on [100]. There are numerous physical medications, for example, plasma treatment, photo oxidation by UV illumination, ionizing radiation, corona, cold plasma, ozone treatment, laser treatment and atmospheric pressure plasma have just been utilized to defeat the inconsistency of different substrates.

2.2.2 Chemical Treatments

The chemical treatment strategy is a notable technique to increase the interfacial quality amongst fibers and matrix. There are different types of chemical treatments have been accounted for in the previous works, have made different levels of progress in enhancing the fiber-matrix bond in green composites [101–105]. The bonding materials and radical induced adhesion upgrade the interfacial strength by delivering covalent bonds between the fiber and the matrix [106]. The adhesion of the fibers with the resin can be enhanced by chemical modifications of the fibers by using   solutions   such   as   alkaline,   iso-cyanate,   silane   permanganate, cyano-ethylation, acetylation, peroxide, vinyl-grafting, organo-silane, benzoly-chloride, styrene, acrylic corrosive and maleic anhydride, maleated polymer, alkoxy-silane, bleaching process, de-waxing process and treated by using other coupling agents. These treatments give various voids on the fiber surface that enhances interlocking between the fiber and the matrix and consequently produces better adhesion [107–109]. The chemical treatments are critical to improve the bonding between the hydrophilic fibers and the hydrophobic matrix at the interface [110–113].

3 Processing of Green Composites

3.1 Woven Fabric Production

Woven fabrics were produced by weaving the fiber yarns utilizing a wooden mould. The square mould of size 400 mm × 400 mm was developed and it contained nails that went about as the twist yarn guider on both of its sides. The weaving procedure was finished by ignoring the weft yarn and underneath the twist yarn, which had been beforehand arranged on the casing with the assistance of the twist yarn guider. The wooden casing, weaving process, and finished woven fabric are appeared in Fig. 3 [114].

3.2 Hatschek Process

The majority of the fabrication methods for green composites are based on the Hatschek process, patented by Hatschek in 1900. It is a semi-continuous process

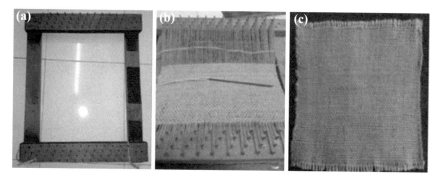

Fig. 3 Woven fabric fabrication **a** wooden frame; **b** weaving process; and **c** woven fabric

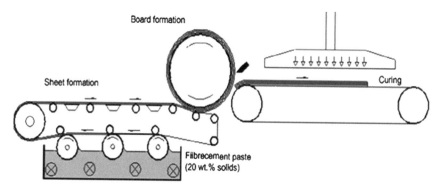

Fig. 4 Scheme of the Hatschek process

comprised of three steps: sheet formation, board formation, and curing. In the first step, a conveyor belt is soaked in a mixture of fresh fiber supplied by a roller from a tank under continuous agitation. Using a vacuum system, a significant portion of the mixing water is removed from the slurry, forming a very thin sheet. The board formation is made in a large cylinder which receives the sheet from the previous step and rolls up in successive layers until the required thickness is achieved. Following this, a guillotine cuts the boards and deposits them on a press to compress and mold the board to the desired shape. Finally, the boards are cured under air or steam which is presented in Fig. 4. These processes produce composites with an adequate percentage of fibers well dispersed into the matrix [13].

3.3 Hand Lay-Up Technique

This technique is one of the most seasoned, easiest and most normally utilized techniques for composite manufacture. The sample is manufactured in layer by layer, and each layer is situated to accomplish the most extreme usage of its

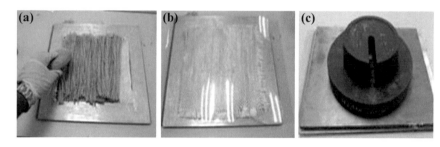

Fig. 5 Stages of green composites fabrication by hand lay-up method [116]

properties. At that point the composites are permitted to cure under atmospheric air conditions and dried under the sun for more than 24 h [9, 115]. The stages of hand lay-up method are illustrated in Fig. 5. At last, the sample was expelled from the mould and dried for 24 h at atmospheric air temperature to confirm that, the sample was hard and sufficiently dry for the testing procedure. At that point the specimen will be cut with the measurement of 100 × 25 mm as indicated by the ASTM benchmarks [116].

3.4 Compression Moulding

The compression or pressure moulding gives a solid bonding between the fiber and matrix [117]. The initial step was to make the material charge by sandwiching the fiber impregnated with matrix. The amount of resin was computed to accomplish the coveted fiber content for each composite. The material charge was then stacked into the mould cavity. Thin parafin sheets of 0.2 mm thickness were utilized as a releasing component to permit a smooth surface finish of the composite plates. Then the mould was then stacked into the furnace and subjected to a compressive load of 75 bar for 15 min to minimal the material charge. The load was then reduced and kept up at 50 bar for 15 min to anticipate resin flash and to limit uneven fiber distribution. The temperature in the furnace and the load were then gradually diminished to surrounding condition and 5 bar respectively for more than 5 min. This prevents the development of voids. The shut mould was permitted to cool under a compressive load of 5 bar for 20 min to inhibit geometrical bending of the composite plate. At last, the fabricated plate was discharged from the mould at ambient temperature [118]. The compression moulding manufacturing assembly is outlined in Fig. 6a. The material load was then stacked into the mould cavity. The detailed view of mould, which is appeared in Fig. 6b, was fabricated by using steel plate.

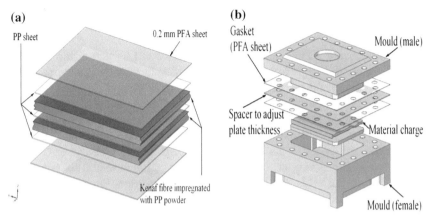

Fig. 6 Compression moulding **a** manufacturing arrangement; **b** mould assembly [118]

3.5 Carding Process

The green composites prepared by carding process is exhibited in Fig. 7. This process gives a uniform mix of the fibers; this is trailed by needle punching, and then pre-squeezing lastly hot squeezing to form the composite material. In this way, this pre-pressed composite was treated with the coupling agent in measures of 1, 3 and 5 parts for every hundred of the pre-pressed composite material. The coupling agent was permitted to infiltrate and pre-react with the pre-pressed material for 2 h. In the last step, the treated pre-pressed material was heated for 5 min at 200 °C under a pressure of 0.7 MPa. This procedure empowered dissolving of the matrix and great impregnation gave an all around combined shaped material [119].

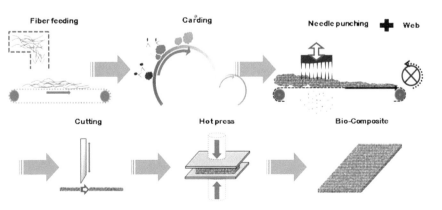

Fig. 7 Carding process for manufacturing of bio-composites

3.6   Mould Method

In this technique the mild steel plates each measures 1 kg and size of 20×20 4 mm^3 are utilized to perform the composite manufacture. Before the procedure starts, clean the mould surface with a release agent to prevent the fiber sticking to the mould and to ease expulsion of the fabricated parts. The part may somehow stick forever to the mould making the composite be rejected. Consequently, the releasing agent was utilized in order to maintain a surface smoothness from the above issue; and to ensure the composite surface is smooth after curing. Then the mould was shut and kept at room temperature for 30 min under the pressure of 0.5 bar to limit the voids. In the wake of curing, composite was isolated from the mould and the samples were cut by the ASTM measurements. Figure 8 demonstrates the arrangement of fibers and matrix in the middle of the mould plate for the manufacture of the composites [120].

3.7   Extrusion Process

The raw materials were set up as per the blending proportions were prepared by utilizing a twin screw extruder. The handling conditions for extrusion were 185, 175, 155 and 140 °C and the screw speed was 150 rpm. The extrudates were chilled off to encompassing temperature in a water shower. At that point the extrudates were cut by utilizing a pelletizer and adequately dried [121]. In another study, the specimens were manufactured by utilizing a co-rotating twin screw extruder. The fibers were at first dried at 50 °C for 24 h before treatment with 3, 6 and 9% NaOH independently. At that point the fibers were washed with water and dried at 60 °C for 24 h. The various components were blended in a high speed rotor blender for 15 min to accomplish homogeneous blending, before dissolve compounding of composites. Then the extrusion was directed at a rotor speed of

Fig. 8 Green composite fabrication process by mould method [120]

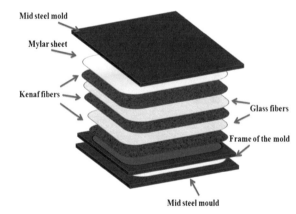

50 rpm. The temperature profile received during compounding was 180 °C at the feed section and expanded to 200 °C at the die head. The extruded fibers were then dried in the atmospheric air and pelletized. Finally, the samples were moulded at 190 °C by utilizing pressure moulding machine [122].

3.8 Pultrusion

In the advancement of fabrication procedures of composites, the strategy which is called pultrusion process has the novel. In addition, pultruded profiles are as of now perceived as an astounding mechanical item, fit for fulfilling an extensive variety of superior and auxiliary component necessities. As to the advantages offered by the pultrusion strategy, the manufacture of composites was first presented and reported by Velde and Kiekens [123]. An examination on the execution of pultruded composites, as a building material for door outlines, has been effectively acquired. They found that pultruded profiles are dimensionally steady and held sufficient mechanical quality under every maturing condition. In addition, an ideal use of coating material has enhanced weathering execution of the pultruded samples, as experienced in outdoor applications. The schematic representation of pultrusion procedure of composite manufacture is demonstrated in Fig. 9 [124].

3.9 Vacuum Infusion Technique

The procedure started by cleaning the surface with acetone to evacuate any earth and adjusted gum from the past infusion process. A thin layer of wax, or impetus, was utilized for the expulsion of the composites after infusion. The fibers were situated on the glass surface took after by a peel employ, netting, and an enka channel, as appeared in Fig. 10. The resin inlet and outlet, which were produced using a plastic pipe, were put over the form territory before it was wrapped with a plastic sheet. The vacuum pump was exchanged on, and the in-shape weight was controlled at below 2000 Pa, to ensure that the air was completely emptied. The resin blend was infused into the mould, where it streamed uniformly until the point

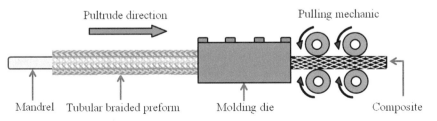

Fig. 9 Pultrusion process of composites manufacture

Fig. 10 Vacuum infusion system

that it achieved the end. The injected composite was expelled from the mould and cured for 24 h at room temperature [114]. The processing methods used by various researchers of green composites are listed in Table 3.

4   Summary and Conclusion

Development in science and innovation add to increase the usage of natural resources and the all inclusive components consider application in automobile, aviation, construction and furniture making, transportation and textile industries. This review presents the research works have been carried out in the field of green composites, focusing on their chemical composition, fiber separation process

Table 3 Processing methods of green composites

Author(s) & year of publication	Name of the journal/proceedings	Type of fiber	Processing method(s)	Refs.
Amash and Zugenmaier (2000)	Polymer	Cellulose	Extrusion	[125]
Angelov et al. (2007)	Composites: part A	Flax	Pultrusion and compression molding	[126]
Arbelaiz et al. (2006)	Thermochim acta	Flax	Injection molding and extrusion	[127]
Arib et al. (2006)	Materials and design	PALF	Film stacking technique	[128]
Shukor et al. (2014)	Materials and design	Kenaf	Compression molding	[122]
Atiqah et al. (2014)	Composites: part B	Kenaf	Mould method	[120]
Doan et al. (2001)	Composites science and technology	Jute	Extrusion and injection molding	[129]
Fung et al. (2003)	Composites science and technology	Sisal	Extrusion and injection molding	[130]
Joseph et al. (2002)	Composites science and technology	Sisal	Extrusion and compression molding	[97]
Khondker et al. (2006)	Composites: part A	Jute	Braiding and compression molding	[131]
Kwon et al. (2014)	Composites: part B	Kenaf	Extrusion process	[121]
Lee et al. (2009)	Composites science and technology	Kenaf	Carding process	[119]
Lee et al. (2007)	Composites: part A	Rice husk and wood flour	Extrusion and injection molding	[132]
Madsen and Lilholt (2003)	Composites science and technology	Flax	Molding method	[133]

(continued)

Table 3 (continued)

Author(s) & year of publication	Name of the journal/proceedings	Type of fiber	Processing method(s)	Refs.
Malkapuram et al. (2009)	Journal of reinforced plastics and composites	Pine wood	Injection molding	[134]
		Flax	Extrusion and injection molding	
		Date palm leaves		
		Sisal, coir, luff sponge	Compression molding	
		Coir	Compression molding	
		Alfa, wood, saw dust	Compression and injection molding	
		Eucalyptus	Hydraulic pressing and injection molding	
		Luffa	Compression molding	
		Lignocel	Injection moulding	
		Aspen pulp	Extrusion and compression molding	
		Hemp	Injection and compression moulding	
		Paper waste	Extrusion and compression molding	
		Rice husk	Extrusion and injection molding	
		Wheat straw	Injection moulding	
Maslinda et al. (2017)	Composite structures	Cellulose	Vacuum infusion technique	[114]
Memona and Nakai (2013)	Energy procedia	Jute	Pultrusion process	[124]
Mohanty et al. (2005)	Natural fibers, biopolymers and biocomposites	Hemp	Compression molding and extrusion	[76]

Table 3 (continued)

Author(s) & year of publication	Name of the journal/proceedings	Type of fiber	Processing method(s)	Refs.
Mwaikambo and Ansell (2002)	Journal of applied polymer science	Kapok and cotton	Hydraulic pressing	[98]
Palanikumar et al. (2016)	Journal of natural fibers	Sisal	Hand lay-up technique	[135]
Pickering et al. (2016)	Composites: part A	Cellulose	Extrusion and compression molding	[136]
Rana et al. (2003)	Composites science and technology	Jute	Injection molding	[137]
Ramesh et al. (2013)	Composites: part B	Sisal and jute	Hand lay-up technique	[9]
Ramesh et al. (2013)	Procedia engineering	Sisal and jute	Hand lay-up technique	[115]
Shahzad (2012)	Journal of composite materials	Kenaf	Extrusion process	[51]
Thwe and Liao (2003)	Composites science and technology	Bamboo	Injection molding	[138]
Velde and Kiekens (2001)	Composite structures	Cellulose	Pultrusion process	[123]
Wambua et al. (2003)	Composites science and technology	Sisal, kenaf, hemp, jute and coir	Compression molding	[139]
Zampaloni et al. (2007)	Composites: part A	Kenaf	Compression molding	[140]
Peng et al. (2011)	Journal of composite materials	Hemp	Pultrusion process	[141]
Bachtiar et al. (2008)	Materials and design	Sugar palm fiber	Hand layup process	[142]
Bachtiar et al. (2009)	Polymer plastics technology and engineering	Sugar palm fiber	Hand layup process	[143]
Ishak et al. (2009)	International journal of mechanical and materials engineering	Sugar palm fiber	Hand layup process	[144]
Leman et al. (2008)	Polymer plastics technology and engineering	Sugar palm fiber	Hand layup process	[145]
Ticoalu et al. (2010)	21st Australasian conference on the mechanics of structures and materials, Melbourne, Australia	Sugar palm fiber	Hand layup process	[146]
Leman et al. (2008)	Materials and design	Sugar palm fiber	Hot press molding	[147]
Sahari et al. (2011)	Key engineering materials	Sugar palm fiber	Compression molding	[148]

and processing methods of composites. The main conclusions of this chapter are as follows:

(i) The demand for natural fibers has seen a marked increment in the most recent decade and experts anticipate a continuation of this pattern in the future.
(ii) Bio-fibers are basically composed by lignin, hemicellulose and cellulose, and the amount, morphology and how these constituents are found in the fibers depend on many factors.
(iii) Among natural fibers, the fibers extracted from the stem of the respective plants are the most common reinforcement materials in polymer composites due to their relatively good specific strength and modulus.
(iv) From the literature, it is found that in addition to environmental friendly nature, their lightness and excellent performance to price ratio contribute to promote the green composites in different industrial applications.
(v) Thus we conclude the chapter that the systematic and persistent research in the future will increase the scope and better future for natural fiber and its green composites.

References

1. Ishak MR, Leman Z, Sapuan SM et al (2013) Chemical composition and FTIR spectra of sugar palm (*Arenga pinnata*) fibers obtained from different heights. J Nat Fiber 10:83–97
2. Meghdad KM, Mortazavi SM (2016) Physical and chemical properties of natural fibers extracted from Typha Australis leaves. J Nat Fiber 13:353–361
3. Faruk O, Bledzki AK, Fink HP et al (2012) Biocomposites reinforced with natural fibers: 2000–2010. Prog Polym Sci 37(11):1552–1596
4. John MJ, Thomas S (2008) Biofibers and biocomposites. Carbohyd Polym 71(3):343–364
5. Madhu P, Sanjay MR, Senthamaraikannan P et al (2017) A review on synthesis and characterization of commercially available natural fibers: part II. J Nat Fiber. https://doi.org/10.1080/15440478.2017.1379045
6. La Mantia FP, Morreale M (2011) Green composites: a brief review. Compos Part A 42(6):579–588
7. Gaceva GB, Avella M, Malinconico M et al (2008) Natural fiber eco-composites. Polym Compos 28(1):98–107
8. Mohanty AK, Misra M, Hinrichsen G (2000) Biofibers, biodegradable polymers and biocomposites: an overview. Macromol Mater Eng 276–277(1):1–24
9. Ramesh M, Palanikumar K, Reddy KH (2013) Mechanical property evaluation of sisal-jute-glass fiber reinforced polyester composites. Compos Part B-Eng 48:1–9
10. Akin DE, Foulk JA, Dodd RB et al (2006) Enzyme-retted flax using different formulations and processed through the USDA flax fiber pilot plant. J Nat Fiber 3:55–68
11. Biagiotti J, Puglia D, Kenny JM (2004) A review on natural fiber-based composites-part I. J Nat Fiber 1(2):37–68
12. Ahmad F, Choi HS, Park MK (2015) A review: natural fiber composites selection in view of mechanical, light weight, and economic properties. Macromol Mater Eng 300(1):10–24
13. Ardanuy M, Claramunt J, Filho RDT (2015) Cellulosic fiber reinforced cement based composites: a review of recent research. Construct Build Mater 79:115–128

14. Faruk O, Bledzki AK, Fink HP et al (2014) Progress report on natural fiber reinforced composites. Macromol Mater Eng 299(1):9–26
15. Gurunathan T, Mohanty S, Nayak SK (2015) A review of the recent developments in biocomposites based on natural fibers and their application perspectives. Compos Part A 77:1–25
16. Nirmal U, Hashim J, Ahmad MMHM (2015) A review on tribological performance of natural fiber polymeric composites. Tribol Int 83:77–104
17. Kamath SS, Sampathkumar D, Bennehalli B (2017) A review on natural areca fibre reinforced polymer composite materials. Ciencia Tecnol dos Mater 29:106–128
18. Ramamoorthy SK, Skrifvars M, Persson A (2015) A review of natural fibers used in biocomposites: plant, animal and regenerated cellulose fibers. Polym Rev 55:107–162
19. Ramesh M (2016) Kenaf (*Hibiscus cannabinus* L.) fiber based bio-materials: a review on processing and properties. Prog Mater Sci 78–79:1–92
20. Sathishkumar TP, Naveen J, Satheeshkumar S (2014) Hybrid fiber reinforced polymer composites—a review. J Reinf Plast Compos 33(5):454–471
21. Thakur VK, Thakur MK (2014) Processing and characterization of natural cellulose fibers/thermoset polymer composites. Carbohyd Polym 109:102–117
22. Thakur VK, Thakur MK, Gupta RK (2014) Review: raw natural fiber based polymer composites. Int J Polym Anal Charact 19(3):256–271
23. Nishino T, Matsuda I, Hirao K (2004) All-cellulose composite. Macromol 37:7683–7687
24. Huber T, Pang S, Staiger MP (2012) All-cellulose composite laminates. Compos Part A 43:1738–1745
25. Kalka S, Huber T, Steinberg J et al (2014) Biodegradability of all-cellulose composite laminates. Compos Part A 59:37–44
26. Ghavami K, Filho RDT, Barbosa NP (1999) Behavior of composite soil reinforced with natural fibers. Cem Concr Compos 21(1):39–48
27. Hyness NRJ, Vignesh NJ, Senthamaraikannan P et al (2017) Characterization of new natural cellulosic fiber from *Heteropogon contortus* plant. J Nat Fiber. https://doi.org/10.1080/15440478.2017.1321516
28. Manimaran P, Senthamaraikannan P, Murugananthan K et al (2017) Physicochemical properties of new cellulosic fibers from *Azadirachta indica* plant. J Nat Fiber. https://doi.org/10.1080/15440478.2017.1302388
29. Reis J (2006) Fracture and flexural characterization of natural fiber-reinforced polymer concrete. Construct Build Mater 20(9):673–678
30. Silva FA, Filho RDT, Fairbairn EMR (2010) Physical and mechanical properties of durable sisal fiber cement composites. Construct Build Mater 24(5):777–785
31. Zhu HX, Yan LB, Zhang R et al (2012) Size-dependent and tunable elastic properties of hierarchical honeycombs with regular square and equilateral triangular cells. Acta Mater 60(12):4927–4939
32. Arpitha G, Sanjay MR, Senthamaraikannan P et al (2017) Hybridization effect of sisal/glass/epoxy/filler based woven fabric reinforced composites. Exp Tech. https://doi.org/10.1007/s40799-017-0203-4
33. Coutts RSP (2005) A review of Australian research into natural fiber cement composites. Cem Concr Compos 27:518–526
34. Kim NK, Lin RJT, Bhattacharyya D (2017) Flammability and mechanical behaviour of polypropylene composites filled with cellulose and protein based fibers: a comparative study. Compos Part A 100:215–226
35. Ku H, Wang H, Pattarachaiyakoop N et al (2011) A review on the tensile properties of natural fiber reinforced polymer composites. Compos Part B 42:856–873
36. Lu TJ, Jiang M, Jiang ZG et al (2013) Effect of surface modification of bamboo cellulose fibers on mechanical properties of cellulose/epoxy composites. Compos Part B Eng 51:28–34

37. Maheshwaran MV, Hyness NRJ, Senthamaraikannan P et al (2017) Characterization of natural cellulosic fiber from *Epipremnum aureum* stem. J Nat Fiber. https://doi.org/10.1080/15440478.2017.1364205
38. Ramesh M, Palanikumar K, Reddy KH (2017) Plant fiber based bio-composites: sustainable and renewable green materials. Renew Sustain Ener Rev 79:558–584
39. Sanjay MR, Yogesha B (2017) Studies on hybridization effect of jute/kenaf/E-glass woven fabric epoxy composites for potential applications: effect of laminate stacking sequences. J Ind Text. https://doi.org/10.1177/1528083717710713
40. Tonoli GHD, Belgacem MN, Siqueira G et al (2013) Processing and dimensional changes of cement based composites reinforced with surface-treated cellulose fibers. Cem Concr Compos 37:68–75
41. Mohanty AK, Misra M, Drzal LT (2002) Sustainable bio-composite from renewable resources: opportunities and challenges in the green materials world. J Polym Environ 10:19–26
42. Netravali AN, Chabba S (2003) Composites get greener. Mater Today 6:22–29
43. Ramesh M, Palanikumar K, Reddy KH (2014) Impact behaviour analysis of sisal/jute and glass fiber reinforced hybrid composites. Adv Mater Res 984–985:266–272
44. Obi Reddy C, Umamaheswari E, Muzenda M et al (2016) Extraction and characterization of cellulose from pretreated ficus (peepal tree) leaf fibers. J Nat Fiber 13:54–64
45. Samson R, Tomkova B (2015) Morphological, thermal, and mechanical characterization of *Sansevieria trifasciata* fibers. J Nat Fiber 12:201–210
46. Kumar RN, Hynes RJ, Senthamaraikannan P et al (2017) Physicochemical and thermal properties of *Ceiba pentandra* bark fiber. J Nat Fiber. https://doi.org/10.1080/15440478.2017.1369208
47. Akil HM, Omar MF, Mazuki AAM et al (2011) Kenaf fiber reinforced composites: a review. Mater Des 32:4107–4121
48. Li Y, Mai YW, Ye L (2000) Sisal fiber and its composites: a review of recent developments. Compos Sci Technol 60(11):2037–2055
49. Lobovikov M, Paudel S, Piazza M et al (2007) World bamboo resources: a thematic study prepared in the framework of the global forest resources assessment 2005. FAO, Rome, pp 1–37
50. Mohanty AK, Misra M (1995) Studies on jute composites: a literature review. Polym Plast Technol Eng 34(5):729–792
51. Shahzad A (2012) Hemp fiber and its composites: a review. J Compos Mater 46:973–986
52. Staiger MP and Tucker (2008) Natural fiber composites in structural applications. In: Properties and performance of natural-fiber composites. Woodhead Publishing, UK, pp 269–300
53. Yan L, Chouw N, Jayaraman K (2014) Effect of triggering and polyurethane foam filler on axial crushing of natural flax/epoxy composite tubes. Mater Des 56:528–541
54. Yan L, Kasal B, Huang L (2016) A review of recent research on the use of cellulosic fibers, their fiber fabric reinforced cementitious, geo-polymer and polymer composites in civil engineering. Compos B 92:94–132
55. Yussuf AA, Massoumi I, Hassan A (2010) Comparison of polylactic acid/kenaf and polylactic acid/rise husk composites: the influence of the natural fibers on the mechanical, thermal and biodegradability properties. J Polym Environ 18:422–429
56. Satyanarayana KG, Sukumaran K, Mukherjee PS et al (1990) Natural fiber–polymer composite. Cement Compos 12:117–136
57. Munawar SS, Umemura K, Kawai S (2007) Characterization of the morphological, physical, and mechanical properties of seven nonwood plant fiber bundles. J Wood Sci 53:108–113
58. John MJ, Anandjiwala RD (2008) Recent developments in chemical modification and characterization of natural fiber-reinforced composites. Polym Compos 29:187–207
59. Mylsamy K, Rajendran I (2010) Investigation on physio-chemical and mechanical properties of raw and Alkali-treated agave Americana fiber. J Reinf Plast Compos 29:2925–2935

60. Ali M (2009) Natural fibers as construction materials. Non-conventional materials and technologies. In: Proceedings of the 11th international conference on non-conventional materials and technologies, Bath, UK
61. Rao KMM, Rao KM (2007) Extraction and tensile properties of natural fibers: vakka, date and bamboo. Compos Struct 77:288–295
62. Joseph K, Toledo RD, James B et al (1999) A review on sisal fiber reinforced polymer composites. Revista Brasileira de Engenharia Agricola e Ambiental 3:367–379
63. Obi Reddy K, Sivamohan Reddy G, Uma Maheswari C et al (2010) Structural characterization of coconut tree leaf sheath fiber reinforcement. J Forestry Res 21:53–58
64. Monteiro SN, Aquino RCMP, Lopes FPD (2008) Performance of curaua fibers in pullout tests. J Mater Sci 43:489–493
65. Rao KMM, Prasad AVR, Babu MNVR et al (2007) Tensile properties of elephant grass fiber reinforced polyester composites. J Mater Sci 42:3266–3272
66. Beckwith SW (2008) Natural fibers: nature providing technology for composites. SAMPE J 44:64–65
67. Reddy GV, Naidu SV, Shobharani T (2009) A study on hardness and flexural properties of kapok/sisal composites. J Reinf Plast Compos 28:2035–2044
68. De Rosa IM, Kenny JM, Mohd M et al (2011) Effect of chemical treatments on the mechanical and thermal behavior of okra (*Abelmoschus esculentus*) fibers. Compos Sci Technol 71:246–254
69. Venkateshwaran N, Elayaperumal A (2010) Banana fiber reinforced polymer composites—a review. J Reinf Plast Compos 29:2387–2396
70. De Rosa IM, Santulli C, Sarasini F (2010) Mechanical and thermal characterization of epoxy composites reinforced with random and quasi-unidirectional untreated *Phormium tenax* leaf fibers. Mater Des 31:2397–2405
71. Sreenivasan VS, Somasundaram S, Ravindran D et al (2011) Microstructural, physico-chemical and mechanical characterization of *Sansevieria cylindrica* fibers—an exploratory investigation. Mater Des 32:453–461
72. Sathishkumar TP, Navaneethakrishnan P, Shankar S (2012) Tensile and flexural properties of snake grass natural fiber reinforced isophthallic polyester composites. Compos Sci Technol 72:1183–1190
73. Sathishkumar TP, Navaneethakrishnan P, Shankar S et al Mechanical properties of randomly oriented snake grass fiber with banana and coir fiber-reinforced hybrid composites. J Compos Mater. https://doi.org/10.1177/0021998312454903
74. Davies P, Morvan C, Sire O et al (2007) Structure and properties of fibers from sea-grass (*Zostera marina*). J Mater Sci 42:4850–4857
75. Le Digabel F, Averous L (2006) Effect of lignin content on the properties of lignocellulose-based biocomposites. Carbohyd Polym 66:537–545
76. Mohanty AK, Misra M and Drzal LT (2005) Natural fibers, biopolymers and biocomposites. CRC Press, pp 1–36
77. Sanjay MR, Arpitha GR, Naik LL et al (2016) Applications of natural fibres and its composites: an overview. Nat Resour 7:108–114
78. Kadolph SJ, Langford AL (2001) Textiles, 9th edn. Prentice Hall, Upper Saddle River
79. Bongarde US, Shinde VD (2014) Review on natural fibre reinforcement polymer composites. Int J Eng Sci Innov Technol 3(2):431–436
80. Mussig J (2010) Industrial applications of natural fibres: structure, properties and technical applications. Wiley, New Jersey
81. Kovacevic Z, Vukusic SB, Zimniewska M (2012) Comparison of Spanish broom (*Spartium junceum* L.) and flax (*Linum usitatissimum*) fibre. Text Res J 82(17):1786–1798
82. de Albuquerque AC, Joseph K, de Carvalho LH et al (2000) Effect of wettability and ageing conditions on the physical and mechanical properties of uniaxially oriented jute-roving-reinforced polyester composites. Compos Sci Technol 60:833–844

83. Ramesh M, Deepa C, Aswin US et al (2017) Effect of alkalization on mechanical and moisture absorption properties of *Azadirachta indica* (Neem Tree) fiber reinforced green composites. Trans Indian Inst Met 70(1):187–199

84. Barreto ACH, Rosa DS, Fechine PBA et al (2011) Properties of sisal fibers treated by alkali solution and their application into cardanol-based biocomposites. Compos Part A 42:492–500

85. Campos A, Marconcini JM, Franchetti SMM et al (2012) The influence of UV-C irradiation on the properties of thermoplastic starch and polycaprolactone biocomposite with sisal bleached fibers. Polym Degrad Stab 97(10):1948–1955

86. Kabir MM, Wang H, Aravinthan T et al (2011) Effects of natural fiber surface on composite properties: a review. eddBE Proc Ener Environ Sustain, pp 94–99

87. Milanese, Ceclia A, Cioffi H et al (2011) Mechanical behavior of natural fiber composites. Proc Eng 10:2022–2027

88. Sulawan K, Wimonlak S, Kasama J (2010) Effect of heat treated sisal fiber on physical properties of polypropylene composites. Adv Mater Res 123–125:1123–1126

89. Li X, Tabil LG, Panigrahi S (2007) Chemical treatments of natural fiber for use in natural fiber-reinforced composites: a review. J Polym Environ 15:25–33

90. Mohanty AK, Misra M, Drzal LT (2001) Surface modifications of natural fibers and performance of the resulting biocomposites: an overview. Compos Interf 8(5):313–343

91. Scarponi C, Pizzinelli CS (2009) Interface and mechanical properties of natural fibers reinforced composites: A review. Int J Mater Prod Technol 36:278–303

92. Almeida AEFS, Tonoli GHD, Santos SF et al (2013) Improved durability of vegetable fiber reinforced cement composite subject to accelerated carbonation at early age. Cem Concr Compos 42:49–58

93. Filho RDT, Silva FA, Fairbairn EMR et al (2009) Durability of compression molded sisal fiber reinforced mortar laminates. Construct Build Mater 23(6):2409–2420

94. Wei J, Meyer C (2015) Degradation mechanisms of natural fiber in the matrix of cement composites. Cem Concr Res 73:1–16

95. Bhoopathi R, Deepa C, Sasikala G et al (2015) Experimental investigation on mechanical properties of hemp-banana-glass fiber reinforced composites. Appl Mech Mater 766–767:167–172

96. Joseph K, Thomas S, Pavithran C (1996) Effect of chemical treatment on the tensile properties of short sisal fiber-reinforced polyethylene composites. Polymer 37:5139–5149

97. Joseph P, Rabello MS, Mattoso LH et al (2002) Environmental effects on the degradation behaviour of sisal fiber reinforced polypropylene composites. Compos Sci Technol 62:1357–1372

98. Mwaikambo LY, Ansell MP (2002) Chemical modification of hemp, sisal, jute, and kapok fibers by alkalization. J Appl Polym Sci 84:2222–2234

99. Ramesh M, Elvin RP, Palanikumar K et al (2011) Surface roughness optimization of machining parameters in machining of composite materials. Int J Appl Res Mech Eng 1(1):26–32

100. Gon D, Das K, Paul P et al (2012) Jute composites as wood substitute. Int J Tex Sci 1(6):84–93

101. Mohanty AK, Khan MA, Hinrichsen G (2000) Effect of chemical modification on the performance of biodegradable jute yarn biopol composites. J Mater Sci 35:2589–2595

102. Mohanty AK, Khan MA, Hinrichsen G (2000) Influence of chemical surface modification on the properties of biodegradable jute fabrics-polyester amide composites. Compos Part A 31(2):143–150

103. Sanadi AR, Caulfield DF (2000) Transcrystalline interphases in natural fiber-PP composites: effect of coupling agent. Compos Interf 7(1):31–43

104. Nunna S, Chandra PR, Shrivastava S et al (2012) A review on mechanical behavior of natural fiber based hybrid composites. J Reinf Plast Compos 31(11):759–769

105. Ramesh P, Prasad BD, Narayana KL (2018) Characterization of kenaf fiber and its composites: a review. J Reinf Plast Compos. https://doi.org/10.1177/0731684418760206

106. Brahmakumar M, Pavithran C, Pillai RM (2005) Coconut fiber reinforced polyethylene composites: effect of natural waxy surface layer of the fiber on fiber/matrix interfacial bonding and strength of composites. Compos Sci Technol 65:563–569
107. Yousif BF, Ku H (2012) Suitability of using coir fiber/polymeric composite for the design of liquid storage tanks. Mater Des 36:847–853
108. Yousif BF, Tayeb NE (2008) Adhesive wear performance of T-OPRP and UT-OPRP composites. Tribol Lett 32(3):199–208
109. Yousif BF (2009) Frictional and wear performance of polyester composites based on coir fibers. In: Proceedings of the institution of mechanical engineers, Part J. J Eng Tribol 223:51–59
110. Hong CK, Hwang I, Kim N et al (2008) Mechanical properties of silanized jute-polypropylene composites. J Ind Eng Chem 14:71–76
111. John MJ, Francis B, Varughese KT et al (2008) Effect of chemical modification on properties of hybrid fiber biocomposites. Compos Part A 39:352–363
112. Taib RM, Ramarad S, Ishak ZAM et al (2009) Effect of immersion time in water on the tensile properties of acetylated steam-exploded *Acacia mangium* fibers filled polyethylene composites. J Thermoplast Compos Mater 22:83–98
113. Khalil HPSA, Tehrani MA, Davoudpour Y et al (2013) Natural fiber reinforced poly(vinyl chloride) composites: a review. J Reinf Plast Compos 32(5):330–356
114. Maslinda AB, Majid MSA, Ridzuan MJM et al (2017) Effect of water absorption on the mechanical properties of hybrid interwoven cellulosic-cellulosic fiber reinforced epoxy composites. Compos Struct 167:227–237
115. Ramesh M, Palanikumar K, Reddy KH (2013) Comparative evaluation on properties of hybrid glass fiber-sisal/jute reinforced epoxy composites. Proc Eng 51:745–750
116. Ghani MAA, Salleh Z, Hyie KM et al (2012) Mechanical properties of kenaf/fiber glass polyester hybrid composite. Proc Eng 41:1654–1659
117. Bernard M, Khalina A, Ali A et al (2011) The effect of processing parameters on the mechanical properties of kenaf fiber plastic composite. Mater Des 32:1039–1043
118. Asumani OML, Reid RG, Paskaramoorthy R (2012) The effects of alkali-silane treatment on the tensile and flexural properties of short fiber non-woven kenaf reinforced polypropylene composites. Compos Part A 43:1431–1440
119. Lee BH, Kim HS, Lee S et al (2009) Bio-composites of kenaf fibers in polylactide: role of improved interfacial adhesion in the carding process. Compos Sci Technol 69:2573–2579
120. Atiqah A, Maleque MA, Jawaid M et al (2014) Development of kenaf-glass reinforced unsaturated polyester hybrid composite for structural applications. Compos Part B 56:68–73
121. Kwon HJ, Sunthornvarabhas J, Park JW et al (2014) Tensile properties of kenaf fiber and corn husk flour reinforced poly(lactic acid) hybrid bio-composites: role of aspect ratio of natural fibers. Compos Part B 56:232–237
122. Shukor F, Hassan A, Islam MS et al (2014) Effect of ammonium polyphosphate on flame retardancy, thermal stability and mechanical properties of alkali treated kenaf fiber filled PLA bio-composites. Mater Des 54:425–429
123. Velde KV, Kiekens P (2001) Thermoplastic pultrusion of natural fiber reinforced composites. Compos Struct 54:355–360
124. Memona A, Nakai A (2013) Mechanical properties of jute spun yarn/PLA tubular braided composite by pultrusion molding. Ener Proc 34:818–829
125. Amash A, Zugenmaier P (2000) Morphology and properties of isotropic and oriented samples of cellulose fiber-polypropylene composites. Polym 41:1589–1596
126. Angelov I, Wiedmer S, Evstatiev M et al (2007) Pultrusion of a flax/polypropylene yarn. Compos Part A-Appl Sci Manuf 38(5):1431–1438
127. Arbelaiz A, Fernandez B, Ramos JA et al (2006) Thermal and crystallization studies of short flax fiber reinforced polypropylene matrix composites: effect of treatments. Thermochim Acta 440:111–121
128. Arib RMN, Sapuan SM, Ahmad MMHM et al (2006) Mechanical properties of pineapple leaf fiber reinforced polypropylene composites. Mater Des 27:391–396

129. Doan TTL, Brodowsky H, Mader E (2001) Jute fiber/polypropylene composites II: Thermal, hydrothermal and dynamic mechanical behavior. Composi Sci Technol 67:2707–2714
130. Fung KL, Xing XS, Li RKY et al (2003) An investigation on the processing of sisal fiber reinforced polypropylene composites. Compos Sci Technol 63:1255–1258
131. Khondker OA, Ishiaku US, Nakai A et al (2006) A novel processing technique for thermoplastic manufacturing of unidirectional composites reinforced with jute yarns. Compos Part A-Appl Sci 37(12):2274–2284
132. Lee SH, Wang S, George P et al (2007) Evaluation of interphase properties in a cellulose fiber-reinforced polypropylene composite by nano-indentation and finite element analysis. Compos Part A-Appl Sci Manuf 38(6):1517–1524
133. Madsen B, Lilholt H (2003) Physical and mechanical properties of unidirectional plant fiber composites: an evaluation of the influence of porosity. Compos Sci Technol 63:1265–1272
134. Malkapuram R, Kumaran V, Negi YS (2009) Recent development in natural fiber reinforced polypropylene composites. J Reinf Plast Compos 28(10):1169–1189
135. Palanikumar K, Ramesh M, Reddy KH (2016) Experimental investigation on the mechanical properties of green hybrid sisal and glass fiber reinforced polymer composites. J Nat Fiber 13(3):321–331
136. Pickering KL, Efendy MGA, Le TM (2016) A review of recent developments in natural fiber composites and their mechanical performance. Compos A 83:98–112
137. Rana AK, Mandala A, Bandyopadhyay S (2003) Short jute fiber reinforced polypropylene composites: effect of compatibiliser, impact modifier and fiber loading. Compos Sci Technol 63:801–806
138. Thwe MM, Liao K (2003) Durability of bamboo-glass fiber reinforced polymer matrix hybrid composites. Compos Sci Technol 63:375–387
139. Wambua P, Ivens J, Verpoest I (2003) Natural fibers: can they replace glass in fiber reinforced plastics. Compos Sci Technol 63:1259–1264
140. Zampaloni M, Pourboghrat F, Yankovich SA et al (2007) Kenaf natural fiber reinforced polypropylene composites: a discussion on manufacturing problems and solutions. Compos Part A 38(6):1569–1580
141. Peng X, Fan M, Hartley J et al (2011) Properties of natural fiber composites made by pultrusion process. J Compos Mater 46(2):237–246
142. Bachtiar D, Sapuan SM, Hamdan MM (2008) The effect of alkaline treatment on tensile properties of sugar palm fibre reinforced epoxy composites. Mater Des 29:1285–1290
143. Bachtiar D, Sapuan SM, Hamdan MM (2009) The influence of alkaline surface fibre treatment on the impact properties of sugar palm fibre-reinforced epoxy composites. Polym Plast Technol Eng 48:379–383
144. Ishak MR, Leman Z, Sapuan SM et al (2009) The effect of sea water treatment on the impact and flexural strength of sugar palm fibre reinforced epoxy composites. Int J Mech Mater Eng 4:316–320
145. Leman Z, Sapuan SM, Azwan M et al (2008) The effect of environmental treatments on fiber surface properties and tensile strength of sugar palm fiber-reinforced epoxy composites. Polym Plast Technol Eng 47:606–612
146. Ticoalu A, Aravinthan T, Cardona F (2010) Experimental investigation into gomuti fibres/ polyester composites. In: Fragomeni S, Venkatesan S, Lam NTK and Setunge S (eds) 21st Australasian conference on the mechanics of structures and materials, Melbourne, Australia. The Netherlands: CRC Press/Balkema, 7–10 Dec 2010, pp. 451–456
147. Leman Z, Sapuan SM, Saifol AM et al (2008) Moisture absorption behavior of sugar palm fiber reinforced epoxy composites. Mater Des 29:1666–1670
148. Sahari J, Sapuan SM, Ismarrubie ZN et al (2011) Comparative study of physical properties based on different parts of sugar palm fibre reinforced unsaturated polyester composites. Key Eng Mater 471–472:455–460

Sisal Fibers Reinforced Epoxidized Nonedible Oils Based Epoxy Green Composites and Its Potential Applications

Sushanta K. Sahoo, Vinay Khandelwal and Gaurav Manik

Abstract Renewable resourced polymer composites from vegetable oils and bio-fibers are receiving increasing attention from various industries due to their characteristics of being less heavy, environment friendly, and biodegradable. Lignocellulosic natural fibers have immense potential to be used as reinforcing fillers due to their characteristics of being less expensive, abundant obtainability, lower density, higher specific strength and modulus, and good interfacial strength with thermoset polymers. In this chapter, epoxidized nonedible linseed and castor oils are proposed as a diluent to petro-based epoxy in formulating toughened bio-based copolymers. Unidirectional sisal fibers were reinforced within a network of such bio-epoxy copolymers in order to achieve an optimal stiffness–toughness balance. Cardanol based phenalkamine, a bio-renewable crosslinker, is used to develop well toughened sustainable and green composite materials. The composites were subjected to various thermal, mechanical, dynamic mechanical, and morphological tests to investigate the impact of nonedible epoxidized oils and sisal fibers in addition to the petro-based epoxy matrix. The present study shows the method for design and development of novel sustainable green composites with higher bio-source content (>65%) meant for shock absorbing applications. These green materials may find good space in making high-performance engineering applications in automotive, structural, construction, and building sectors.

Keywords Renewable resources · Epoxidized oils · Bio-fibers
Green composites · Mechanical properties

S. K. Sahoo · V. Khandelwal · G. Manik (✉)
Department of Polymer and Process Engineering, Indian Institute of Technology Roorkee, Uttarakhand, India
e-mail: manikfpt@iitr.ac.in

© Springer Nature Singapore Pte Ltd. 2019
S. S. Muthu (ed.), *Green Composites*, Textile Science and Clothing Technology,
https://doi.org/10.1007/978-981-13-1972-3_3

73

1 Introduction

Diminishing petro-based resources and environmental apprehensions have inspired the researchers to develop sustainable materials which are of renewable origin. Most of the engineering polymers used in various industrial applications are non-renewable resourced and have serious concerns of being hazardous, non-biodegradable, expensive, involve high energy consumption during synthesis, and create an unhealthy industrial environment [1–5]. Thus, there is growing interest to synthesize greener materials from renewable feedstocks to use in coating, adhesive and composite industries. Partial or complete replacement of petro-based resins with bio-based resins would improve the physical properties of reconstituted materials [3, 4]. In addition to that, such bio-resins reduce the carbon dioxide footprint and have no adverse effect on health and environment. Further, replacement of petro-based resins with bio-based resins would improve the physical properties of reconstituted materials. Consequently, bio-based polymers and composites have drawn significant attention in both academic and industries.

  The resins derived from biological resources are known as bio-resins which must be green and sustainable. The bio-resins can compete with less sustainable technologies being economical, nontoxic, and benign nature in production and in use to enhance the environmental outline of the products. The bio-resins are used as binders or diluent to improve the performance of polymers by replacing existing technologies. Most importantly, the properties are more or less comparable with synthetic petroleum-based resin. There are so many types of bio-resins based on their origin and structure such as Tanin, Lignin, Furan, Rosin, CNSL, Carbohydrates, Itanoic acid, Proteins, etc. [2–6]. Bio-resins derived from these resources have been the center of attraction in the field of research as an alternative to epoxy for sustainable development of bio-based thermoset polymers. However, some of these epoxies still have serious drawbacks such as limited production, low purity, complex structure, lack of structural control, hydrophilicity, brittleness, and unknown toxicity. The effective and low cost production of a bio-sourced epoxy with higher performance has remained a challenge [6–8]. In this context, among all renewable-based resources, vegetable oil sourced bio-resins are widely used as a precursor to developing bio-based polymers because of their several benefits such as abundant availability, inexpensiveness, bio-renewability, environment friendliness, and presence of reacting sites [7]. In recent years, plant oils like soybean, linseed, cottonseed, and castor oils have encouraged the researchers as a resource for the preparation of resins and pre-polymers that may serve as alternatives to petroleum-based resources. Similarly, the inferior mechanical and thermophysical properties of crosslinked functionalized oils can be improved by reinforcing renewable bio-fibers. This concept further trims down the dependence on fossil resources which have associated negative environmental impacts. Natural fibers may offer significant opportunities in the development of green composites derived from vegetable oils with improved performance compared to pure polymers.

In this chapter, we reviewed the literature covering reports on bio-resins, bio-thermoset blends, and lignocellulosic fiber reinforced bio-based thermoset composites. We also present an interesting case study of the synthesis and characterization of epoxidized plant oil based epoxy composites reinforced with sisal fibers.

2 Industrial Application Prospective

Bio-based resins and composites have been widely used in coating, automotive, and structural applications. More efforts are needed to form a novel and improved pre-polymers or polymers in industries using renewable feedstocks. Further, bio-resourced polymers may have inferior mechanical and thermal properties compared to petro-based plastics, which restrict their wider exploration and applications [3, 4]. However, properties of such materials can be improved by either through functionalization or blending with petro-based polymers, and also, incorporating various reinforcing agents to ensure sustainability and environmental protection. For example, supersap bio-epoxy resin with a portion of bio-sourced content is commercially available having acceptable properties suitable for wood coating application.

Similarly, bio-composites based on thermoset resins and natural fibers are already used in making furniture and automotive parts because of their good energy absorbing capacity. Lightweight, inexpensive lignocellulosic fibers have the potential to replace synthetic fibers to be used as suitable reinforcing agents in construction, automotive, and structural applications. In the recent years, European car manufacturing companies like Dieffenbacher, BASF, and Rieter Automotive have used bio-fiber reinforced composites based on thermoplastic and thermoset polymer composites to develop automotive exterior parts like bumpers, seat backs, door panels, package trays, dashboards, etc., and also some interior parts [5]. Cellulosic fibers like banana, hemp, flax, sisal, kenaf, jute, etc., provide reinforcement to matrix along with reduction in weight, low cost, potential for recyclability, green and eco-friendly. Glass fiber reinforced polymer composites meet the specific applications demands with better mechanical performance and has led to a well-built manufacturing base in industries. However, glass-reinforced composites demotivate the industry and environment regulators due to certain shortcomings like high density of fibers, trouble in machining, and difficulty in recycling and the health hazards. Besides automotive industry, other sectors also have qualified an increase in the usage of bio-resins and natural fibers. The utilization of bio-fibers in the structural, building, and construction industries are growing at the rate of 13% in the last decade. Ford motor also carried out the research and developed composites using natural fibers, soy resin, hemp, and sisal to manufacture exterior body panels.

Fig. 1 Automotive components based on lignocellulosic fiber reinforced polymer composites [5]

Similarly, Mercedes Benz, Toyota, General Motors companies also developed bio-based composites for different automotive parts in recent years. The bio-composites developed by Mercedes Benz using fibers is shown in Fig. 1 to refer its possible applications.

3   Plant Oil Based Bio-thermosets

3.1   Plant Oils

Plant oils have received growing attention as a renewable raw material by both the industries and academicians due to their unique advantages such as biodegradability, capability to crosslink, ease of processing, non-harmful nature, and environmentally benign nature, etc. [6]. The life cycle of plant oils based polymers is shown in Fig. 2.

The cycle illustrates that there is no loss of energy or resources after use as biomass obtained from plant oil based polymer waste can be utilized to produce the source oils again. In the last decade, vegetable oils based advanced materials have been developed in the field of thermosetting resins through different processes because of their unsaturation content. Plant oil triglycerides are the esterified glycerol with three long chain fatty acids. The amount of different fatty acids characterizes the oil type. Plant oils mostly contain fatty acids having fourteen to twenty-two carbons in length and one to three double bonds. The characteristics of

Fig. 2 Lifecycle of polymers from plant oils [7]

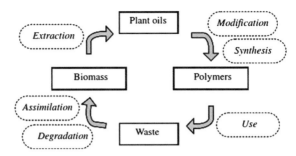

vegetable oils depend on the fatty acid chain and the numbers or position of unsaturated bonds, which render the unique features of the oil. Generally, oleic acid (C18:1), linoleic acid (C18:2), and linolenic acid (C18:3) are the common compositions of the oil as shown in Table 1 [9, 10].

Triglycerides are highly functionalizable molecules because of high unsaturation content, and, therefore, have been used to develop novel crosslinked polymers. The reactive sites or unsaturated bonds in such oils may undergo direct polymerization to form pre-polymers or polymers. Apart from this, the originally present functional groups in triglycerides, such as hydroxyl or epoxide functionality, can be cross-linked using various polymerization methods [9, 11–15]. Most suitable and effective approach to achieve high-performance polymer is the chemical modification of such oils prior to their polymerization. By incorporating easily polymerizable functional groups, the disadvantage of the low reactivity of triglycerides is overcome and the use of plant oils in industries is explored.

Table 1 Availability and composition of vegetable oils [9]

Natural oil	Average annual production [10^6 tons]	Fatty acid					No. of double bonds	Iodine value [mg/ 100 g]
		Oleic	Palmitic	Stearic	Linoleic	Linolenic		
Soybean	26.52	23.4	11.0	4.0	53.3	7.8	4.6	117–143
Palm	23.53	40.5	42.8	4.2	10.1	–	1.7	44–58
Rapeseed	15.29	56	4	4.2	26	10	3.8	94–120
Sunflower	15.29	37.2	5.2	2.7	53.8	1.0	4.7	110–143
Groundnut	5.03	48.3	11.4	2.4	31.9	–	3.4	80–106
Cottonseed	4.49	18.6	21.6	2.6	54.4	0.7	3.9	90–119
Olive	2.52	71.1	13.7	2.5	10.0	0.6	2.8	75–94
Corn	2.30	25.4	10.9	2.0	59.6	1.2	4.5	102–130
Canola	15.28	60.9	4.1	4.2	21.0	8.8	3.9	11–126
Fish	1.13	11–25	10–22	–	–	–	–	104–110
Linseed	0.83	19.1	5.5	3.5	15.3	56.6	6.6	168–204
Sesame	0.76	41	9	6	43	1.0	3.9	103–116
Castor	0.56	5.0	1.5	0.5	4.0	0.5	2.7	82–88

3.2 Chemical Modification of Plant Oils

The plant oil in unmodified form cannot be used as the base matrix in the formation of blends and composites because the unsaturated double bonds present are not sufficient enough to undergo polymerization to provide adequate stiffness and strength [10, 16–18]. These unsaturation or vinyl groups present in the triglyceride chain are incapable for any polymerization reaction apart from direct polymerization, cationic/radical polymerization or oxopolymerization [10, 16–21]. Thus, constant efforts have been undertaken for modification of these groups that can facilitate effective polymerization of triglycerides to fabricate polymers. These modifications include insertion of polymerizable groups into the long glyceride chains, that can render greater ability for undergoing polymerization [22–39]. Thus, constant efforts have been undertaken for modification of these groups to facilitate polymerization of triglycerides to fabricate polymers.

As revealed by various literature reports, the most common approach to obtain high-performance polymers is through the chemical modification of the double bonds in triglyceride by various reactions that can introduce sensitive functional groups, such as hydroxyl, epoxy, acrylate or carboxyl moieties [22–64]. The various reactions which are used for the functionalization of triglycerides include epoxidation, transesterification, and acrylation, maleation. Among these, epoxidation of oils is widely commercialized even at industrial scale due to its versatility in forming various pre-polymers. The epoxidation process is often catalyzed by hydrogen peroxide through chemical or enzymatic oxidation mode which involves reaction around the C=C bonds [25–28]. Similarly, formation of methyl ester groups may result in reducing the viscosity and improving the reactivity of resins which resulted in higher performance. Similarly, the other methodologies are also adapted for the modification of triglycerides which includes the free radical polymerization, ring opening, metathesis polymerization, and polycondensation reactions to form pre-polymers. Recent years have witnessed various synthetic routes for the transformation of vegetable oil to functional polymers by utilizing concepts of sustainable chemistry. The epoxidized derivatives of plant oils are increasingly finding use in numerous applications [65, 66].

3.3 Plant Oil Based Thermoset Blends

Petroleum-based thermosetting polymers have excellent mechanical property, better thermal stability, better adhesion with fibers or other fillers, low expansion coefficient, better dimensional stability, high chemical resistance, lightweight, etc. [67]. Generally, polyesters, epoxies, phenolics and vinyl esters are well-known thermosets and extensively explored in industries owing to their unique properties.

Among all engineering thermoset polymers, epoxy resins are highly used on account of its outstanding mechanical performance, higher thermal degradation stability, little curing shrinkage, and good solvent resistance [68]. The versatility of the epoxy group in undergoing different chemical reactions with a range of chemicals such as polyamines, polyacids, polythiols, polyphenols, etc. enables it to be easily cured without emission of hazardous products. Epoxy, as a matrix, is also chemically compatible with other components like reactive diluents, reinforcing agents, etc., making them suitable for several composite technology applications. Thus, they are normally used as adhesives, coatings, casting materials, potting compounds, etc. The most promising applications are found in the aerospace and automotive industries where resins and fibers are combined to manufacture complex composite structure.

Despite having excellent mechanical and thermal stability, synthetic resins have solemn disadvantages in terms of brittleness or inferior fracture toughness which confined its wide exploration in structural applications. On the other hand, pure bio-resins lack adequate mechanical and thermophysical properties to be used for structural applications. In order to achieve stiffness–toughness balance, oil-derived bio-resins are found to be successful modifiers to toughen the matrix [66]. Further, petroleum-derived resins are blended with the plant oil based resins as a diluent for easy processability, reduced viscosity, improved wettability, and enhanced thermophysical properties. Different particles like rubbers, elastomers, etc. are used as the impact modifiers for polymers, blends, and composites [69]. To perform as a successful modifier, the diluent should have functional groups to be compatible with epoxy resin and have better reactivity. Simultaneously, reduction and controlling the viscosity of base resin matrix is a crucial parameter to improve processibility in the molding process like resin transfer molding, pultrusion, etc. In this context, the oil-based bio-resin modifiers play a vital role in improving processibility and end-use characteristics of material [70]. Functionalized oils can substitute petro-based resins due to chemical modifications with different groups, plentiful availability, low-priced and a wider possibility to take part in crosslinking. Furthermore, it would have a greater positive environmental impact since it would not only reduce the VOC content but also make the material partially bio-based and biodegradable. Many literatures have reported on the toughening of epoxy by addition of epoxidized oil-based bio-resins into epoxy with moderate strength and modulus [65–68, 71–84].

4 Natural Fibers: Potential Applications and Limitations

Synthetic fibers like E-glass fibers and carbon fibers have higher density and are expensive. Thus, the developed composites used as automotive components have higher weight and decreased fuel efficiency. Further, they are nonrenewable and nonbiodegradable raising environmental regulation concerns. Natural fibers plentifully exist in nature and can be utilized as reinforcing agents in polymers to

Table 2 Physical properties of fibers [85]

Fiber	Density (g cm^{-3})	Diameter (µm)	Tensile strength (MPa)	Young's modulus (GPa)	Elongation at break (%)
Flax	1.5	40–600	345–1500	27.6	2.7–3.2
Hemp	1.47	25–500	690	70	1.6
Jute	1.3–1.4	25–200	393–800	13–26.5	1.16–1.5
Kenaf	–	–	930	53	1.6
Ramie	1.55	–	400–938	61.4–128	1.2–3.8
Sisal	1.45	50–200	468–700	9.4–22	3–7
Cotton	1.5–1.6	12–38	287–800	5.5–12.6	7–8
Coir	1.15–1.46	100–460	131–220	4–6	15–40
E-glass	2.55	<17	3400	73	2.5
Kevlar	1.44		3000	60	2.5–3.7
Carbon	1.78	5–7	3400–3800	240–425	1.4–1.8

achieve strong and low-density materials [58–64]. Lignocellulosic natural fibers mostly consist of cellulose, hemicellulose, lignin, and pectin as depicted in Table 2. Cellulose contributes to tensile strength whereas lignin offers thermal stability in bio-fibers. Plant-based bio-fibers are used in commercial applications such as automotive industries, household applications, etc. Several natural fibers such as flax, hemp, jute, and sisal have been used in composite materials to enhance the mechanical performance and additionally increase the bio-based content within composites. Natural fibers have many drawbacks like lack of homogeneity, inferior thermal stability, non-fire resistance, limited compatibility, poor mechanical strength, and higher moisture absorption in comparison to synthetic fibers [85–93]. Some such limitations of natural fibers can be overcome to some extent through chemical modifications but can not be removed completely. However, natural fibers have numerous benefits than synthetic fibers like lower density, inexpensiveness ($0.50/kg), recyclability, biodegradability, good thermal stability, nonabrasive, acoustical insulation properties, and nonhazardous nature which are beneficial for modern industrial applications [85, 94–98]. Further, natural fibers can be grown in a little time, and hence, the agricultural farmers can be motivated to cultivate the fibers as a potential cash crop. Considering the advantages and disadvantages of the composites, these renewable fibers cannot be ignored by the polymer composite industries for automotive, building, construction, and other applications [99].

5 Thermoset Green Composites

The push toward green and sustainable materials in industries have stimulated the researchers to manufacture green composites with both the matrix and fillers obtained from bio-resources. Composites derived from natural resources have immense potential to replace fossil-resourced polymers and are readily acceptable as environment friendly materials. These materials are becoming more significant as the polymers are obtained from renewable feedstocks. The fibers reinforcement within thermoset polymers helps to increase the mechanical and thermophysical properties of the base polymers to make them suitable for specific applications. The natural fiber based thermoset bio-composites are widely explored in aerospace, automotive, and construction industries owing to its higher strength and modulus and improved damping ability [100]. Bio-composites are broadly defined as composites in which either matrix or reinforcement or both components of the system are derived from natural resources. These composites system may be composed of bio-resin and man-made fibers or petro-based resins and natural fibers, or bio-resins and natural fibers. In recent years, the attention toward bio-resourced composites is increasing by industrial researchers owing to their sustainability, eco-friendly nature with higher bio-based content and properties as per technical requirements [86–88, 100–102]. Being the major content of composite, the plant oil based thermoset resins or blends have drawn significant interest to be used as base matrices in manufacturing green and sustainable materials. Less viscous plant oil based resins act as binders to bind the fibers and matrix together so that major loads can be transferred to the fibers. Although, glass, aramid, and carbon fibers are broadly used as reinforcing agents in polymer composites for tremendous improvement in properties, however, they have high density and also sourced from nonrenewable origins. In this context, bio-composites based on plant oil based resins and natural fibers offer several advantages including low density, superior mechanical performance, better processability, and fair impregnation with fibers, cost-effective, partially or completely biodegradability [103]. Additionally, bio-fibers reduced the chance of tool wear during processing and have no adverse health effects on workers in industries. Since, crosslinked plant oils as a base matrix are unable to offer required strength and modulus to the composites for specific applications; thus, thermoset blends based on plant oils are used as effective matrices to widen the application fields with proper stiffness–toughness balance. In this perspective, plant oil based thermoset green composites pave the way for design and development of sustainable and green composites with higher bio-based content for energy absorbing, and high strength requiring applications.

5.1 Plant Oil Based Thermoset Composites

Plant oil based thermoset polymers contribute to the development of green materials by trimming down the dependence on fossil resources and negative environmental impacts. However, plant oil derived polymers and their blends could not find suitability in structural applications due to their inherently low stiffness and strength [60–64, 67, 71, 72, 104–112]. To overcome these drawbacks, reinforcing agents such as fibers, layered plates, and other particles are incorporated within the bio-based matrix. The strength, stiffness, toughness, heat, and chemical resistance can all be improved by the addition of such fillers. Such fiber reinforced plant oil based composites have attracted huge interest in industrial and academic research because of their unique properties like low weight, high strength to mass ratio, nonabrasive, noncorrosive, and enhanced fracture toughness. In this context, natural fibers offer significant opportunities in the development of vegetable oil based bio-composites derived from renewable resources with improved performance compared to pure polymers [86–93, 113, 114].

The natural fiber reinforced plant oil based bio-composites are broadly used in the automotive and construction industries because of inexpensive, higher mechanical performance, sustainability, and environmental friendly. Most importantly, both bio-based matrix and bio-fibers make the composite green or bio-based and biodegradable in totality. Pure epoxidized plant oil or its blends with high loading of bio-resin are not suitable for structural applications on account of inadequate stiffness and strength. Several authors are frequently using natural fibers like jute, sisal, flax, hemp, kenaf, etc., to increase the properties of plant oil based polymers in recent years [115–125].

5.2 Bio-Based Curing Agents for Green Composites

The crosslinkers play a significant role to achieve the essential properties the crosslinked bio-resins via different polymerization to generate stiff, chemical resistant, mechanically, and thermally stable materials. Widely used petro-derived amines and anhydrides have dangerous issues such as toxicity, hazardous, and non-ecofriendly. Further, use of petroleum-based crosslinking agents also offer poor toughness to thermosets which make the product inappropriate for specific applications. Thus, it is important to explore eco-friendly curing agents rather than derived from renewable resources for producing next-generation bio-based thermosets. Similarly, bio-based curing agents are essential for developing more green and sustainable bio-based polymers through ecological process [126, 127]. Phenalkamines (PKA) synthesized from cardanol are typically used as bio-sourced curatives with the benefit of low-slung temperature curing applications [128]. The presence of benzene in PKA offers chemical/solvent resistance wherein, aliphatic chain yields hydrophobic nature, ductility, and long pot life [127, 129, 130]. A combination of long alkyl chain and aromatic moiety, thus maintain outstanding balance in the design of polymers. Similarly, other bio-based curing agents like carboxylic acids: adipic acid, citric acid, tartaric acid, and bio-based anhydride: glutaric anhydride, bio-based amine, and derivatives are also used recently to crosslink petro and bio-resins with acceptable performance [126].

6 Case Study: Sisal Fiber Reinforced Epoxidized Oil Based Epoxy Composites

6.1 Synthesis and Characterization of Epoxidized Oils

6.1.1 Linseed and Castor Oil Epoxidation

Both the oils (linseed and castor oil) were chemically modified by epoxidation method using CH_3COOH and H_2O_2 as described previously [38, 56, 131]. Epoxidization of each oil was taken in a three-necked flask connected with a stirrer with a hot plate. 1:1 molar ratio CH_3COOH and unsaturation of oil, 2:1 of H_2O_2 and unsaturation of oil, 30% (w/w) H_2O_2 and 25 wt% Seralite SRC-120 (acidic ion exchange resin) were taken for the epoxidation at 60 °C. After the reaction, the seralite catalyst is removed from the mixture. After the completion of the reaction, the oil mixture layer was isolated through separating funnel, neutralized with 1.5 wt % Na_2CO_3 solution, and then dehydrated using $MgSO_4$ followed by filtration. The epoxidation scheme of plant oil is displayed in Fig. 3.

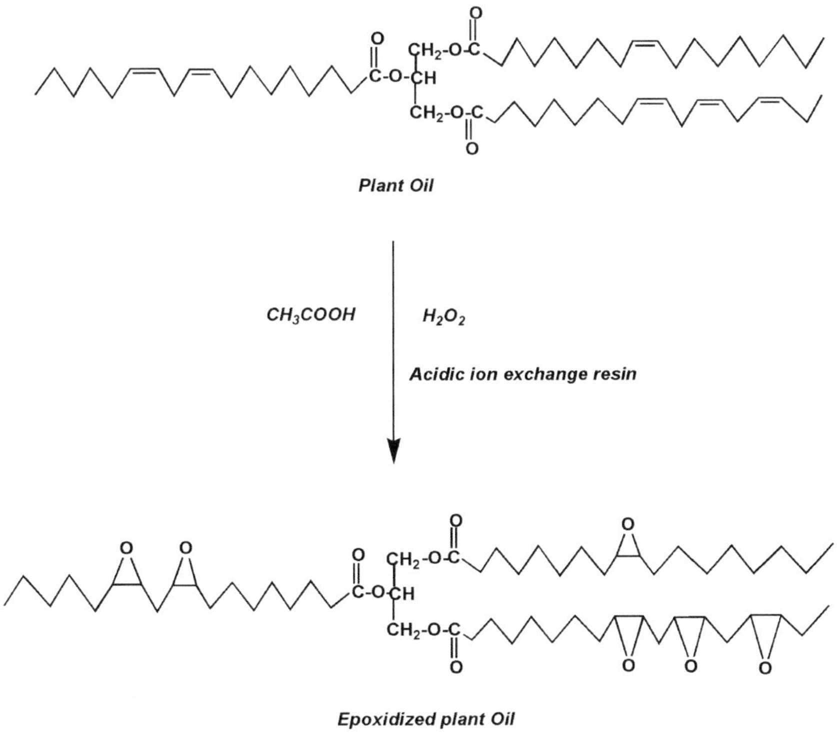

Plant Oil

CH_3COOH H_2O_2

Acidic ion exchange resin

Epoxidized plant Oil

Fig. 3 Scheme of in situ epoxidation of plant oil [38]

6.1.2   Characterization of Epoxidized Oils

Functional Moiety Analysis

The epoxy equivalent weight (EEW) of ELO and ECO was measured as 195 and 320, respectively, through HBr-COOH titration following AOCS Cd-9-57 (1998). Iodine number of linseed oil and castor oil were measured to be 164 and 83, respectively, which decreased to 11 and 6 in that order after functionalization.

FTIR Analysis

Figure 4a shows the IR spectrum of linseed oil and Fig. 4b depicts the spectra of castor oil before and after epoxation. In case of LO, the transmittance signals at 3010 cm^{-1} correspond to –C=C–H and the signals at 1651 cm^{-1} are attributed to unsaturation, –C=C– and –CH=CH–, correspondingly [6] the bands at 3010 and 1651 cm^{-1} disappeared after undergoing epoxation. The bands in the range of 820–843 cm^{-1} ascribed to C–O–C stretch revealed the formation epoxide rings. Similar as reported by Kim and Sharma [38].

Similarly, in the case of castor oil (Fig. 4b), signals noticed at 3009 and 1655 cm^{-1} correspond to unsaturation vibration. The disappearance of these bands and a new band formation at 841 cm^{-1} in ECO confirmed the epoxation of CO as seen in case of LO [56]. Similar investigation was studied previously by Parada Hernandez et al. [35].

NMR Analysis

^{1}H NMR spectra of linseed oil before and after epoxidization are given in Fig. 5a, b correspondingly to reveal the epoxation. In case of LO, peaks at 2.8 ppm is

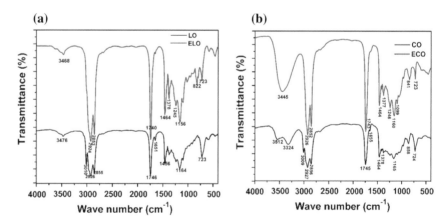

Fig. 4 Assessment of FTIR spectra of **a** LO and ELO and **b** CO and ECO to reveal the epoxation

ascribed to hydrogens attached to unsaturated groups (–CH=CH–CH$_2$–CH=CH–). The multiple peaks at 4.1–4.4 ppm are attributed to methylene protons of (–CH–CH$_2$–O–). Vinyl hydrogens (–CH=CH–) and methiene proton of glycerol moieties (–CH–O–C(O)–) is detected at 5.25–5.42 ppm.

It is seen that the intensity of the peak at 5.4 ppm was significantly reduced after undergoing epoxide formation and transformation of sp^2 to sp^3 hybridization. Further, peak at 2.05 ppm conforming to the allyl protons is shifted to 1.45 ppm afterward the reaction. The –CH– hydrogens of the epoxide group at 2.9–3.12 ppm confirmed the synthesis of ELO. The unsaturation peak almost vanished in the spectra of ELO confirming complete conversion of double bonds to epoxides. Likewise, the same fact has been described for epoxidation of LO using 25% amberlite catalyst earlier [38].

Figure 6a and b depicts proton NMR spectra of castor oil before and after epoxidation, respectively. The peak at 3.64 ppm corresponds to the –OH group of ricinoleic acid. Set of peak signals between 4.1–4.3 ppm accredited to the glycerol moiety (–CH–CH$_2$–O–) as noticed in LO, and the signal between 5.2 and 5.6 ppm is assigned to unsaturated double bonds (–CH=CH–). In Fig. 6b, the peaks at 2.9–3.1 ppm (–CH–O–CH–) characterized the formation of oxirane groups in ricinoleic acid. The signals of hydroxyl noticed at 3.64 ppm prior to epoxidation moved to 3.85 ppm.

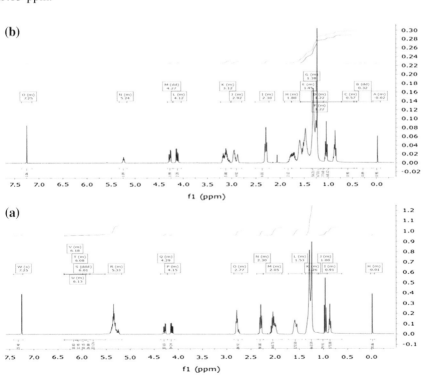

Fig. 5 Proton NMR spectra of **a** LO and **b** ELO

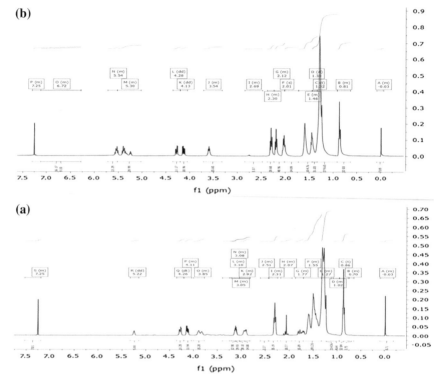

Fig. 6 Proton NMR spectra of **a** CO and **b** ECO

6.2 Formulation of Bio-epoxy Resin Blends

ELO and ECO were used as a secondary component to DGEBA-epoxy at different compositions (10, 20, and 30 wt%) to formulate bio-based epoxy blends. Bio-crosslinker PKA was used in the formulation as per epoxy equivalent weight (EEW) of the resin blends and amine hydrogen equivalent weight (AHEW) of PKA as shown by Eq. (1)

$$\text{PKA phr} = \frac{\text{AHEW} \times 1000}{\text{EEW}} \tag{1}$$

The homogeneous mixture of resins was formulated by stirring the resin mixture at 500 rpm for 30 min, and placed in an oven for some time to eradicate bubbles. After this, calculated PKA curative was poured into the resin blend and, subsequently, the resin blend was poured into a release agent sprayed over steel mold. Epoxy, epoxy with 20% ELO, and epoxy with 30% ELO resins have been coded as EP, EPELO20, and EPELO30, respectively. Similarly, castor oil based resin blends are coded as EPECO10, EPECO20, and EPECO30.

6.3 Manufacturing of Bio-based Epoxy Composite

The prepared resins as mentioned in the previous section were used to manufacture composites. Unidirectional sisal fiber in mat form was used to strengthen the epoxy/ELO and epoxy/ECO matrix. The physical properties of sisal fiber mat and DGEBA-epoxy is depicted in Table 3.

To prepare composite, sisal fiber mat are placed layer by layer in [0/0] pattern with resin mixture rolled over it. Two layers of mat were used and hand lay-up method followed by load was implemented in manufacturing all the composites. For initial curing, the molded samples were placed at 25 °C for one day, and then kept for post curing (120 °C for 2 h followed by 150 °C for 5 h). Complete process and design of the bio-composite product development is shown in Fig. 7 for better understanding.

The developed composites are coded as EPSF, EPELO20SF, and EPELO30SF for epoxy composites with 0, 20, and 30 wt% of ELO bio-resin, respectively. Similarly, EPECO20SF and EPECO30SF are noted for 20 and 30 phr of ECO correspondingly.

6.4 Effect of Epoxidized Oils on Properties of Bio-epoxy Composites

6.4.1 Density and Void Fraction Within Composite

The technique, hand lay-up followed by compression molding is used to prepare all the composites, maintaining the sample's thickness between 4.5–4.8 mm and voids may be present inside the samples. Both the theoretical and experimental densities of all composite samples were evaluated as reported earlier [101] and employed to compute the void fraction as presented in Table 4. It is noticed that with an increase in ELO content, void fraction decreases and density increases. EPELO20SF and EPECO30SF composites show optimal density which could be because of reduced viscosity of resin blends results in better wetting and impregnation of fibers, thereby

Table 3 Physical properties of sisal fiber and DGEBA-epoxy

Property	Epoxy (DGEBA)	Sisal fiber mat
Density at 25 °C (gm/cm^3)	1.21	1.45
Diameter (μm)	–	50–200
EEW (gm/mol)	190	–
Tensile strength (MPa)	50–70	400–600
Elastic modulus (GPa)	2–3	10–13
Strain at break (%)	3–20	2.4–2.9

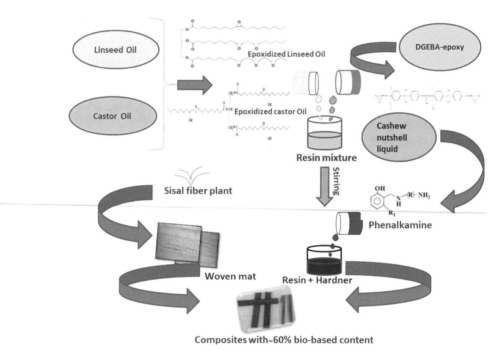

Fig. 7 Design and development process of bio-based epoxy composites

Table 4 Fiber volume fraction, density, and void content of composites

Sample	Fiber volume fraction	Theoretical density (g/cm^3)	Experimental density (g/cm^3)	Void fraction (%)
EPSF	15.93	1.217	1.057	13.22
EPELO20SF	17.98	1.192	1.096	8.05
EPELO30SF	18.45	1.181	1.092	7.53
EPECO20SF	16.98	1.176	1.083	8.29
EPECO30SF	17.37	1.198	1.094	7.86

resulting in the enhanced packing of fibers along with a reduction in voids. In contrast, epoxy unaided is not able to impregnate the fibers due to its high viscosity that leads to loose packing of fibers and creation of cavities or voids resulting in reduced experimental density significantly.

6.4.2 Tensile Properties

The incorporation of a small amount of epoxidized plant oil to epoxy composite enhances the stiffness and strength with strong interfacial interaction as reported

Table 5 Tensile properties of unmodified and modified epoxy bio-composites

Sample	Tensile strength (MPa)	Tensile modulus (GPa)	Elongation at break (%)
EPSF	69 ± 3	2.12 ± 0.06	4.7 ± 0.3
EPELO20SF	85 ± 5	2.30 ± 0.07	6.2 ± 0.2
EPELO30SF	65 ± 3	1.71 ± 0.06	7.1 ± 0.1
EPECO20SF	62 ± 1	2.21 ± 0.03	4.6 ± 0.1
EPECO30SF	54 ± 2	2.10 ± 0.03	5.0 ± 0.1

earlier [36, 132]. Table 5 presents the tensile properties of pure epoxy and epoxilized plant oil modified epoxy composites.

The tensile properties (both the strength and modulus) of EPSF composite raised on adding 20% ELO bio-resin. On the other hand, addition of 20% ECO enhances the stiffness/modulus of the composite along with minor reduction in strength. However, higher amount (30%) of both ELO and ECO decreases the modulus and strength significantly along with higher elongation due to increasing replacement of rigid aromatic groups in petro-epoxy by flexible aliphatic chains. Similar trend in DGEBA/ELO-based composites was investigated earlier by Yim et al. [57] and Sahoo et al. [93]. Likewise, maximum increment in tensile strength for jute fiber reinforced DGEBA-epoxy composite was observed at 10% ESO and 10% EHO as reported by Manthey et al. [123]. In the current work, maximum tensile strength and modulus were observed to be 84.83 MPa and 2.3 GPa for EPELO20SF which is higher than that of EPECO20SF composite. Essentially, 20 phr ELO and 20 phr of ECO reduce the epoxy resin viscosity on addition. This allows the proper wetting or impregnation of fibers that results in their better packing and stronger adhesion at matrix–fiber interface, for which the roughness and the rigidity of composites are increased [92]. Comparatively, higher tensile strength is noticed in ELO-modified epoxy composites because of similar epoxy value of ELO as that of DGEBA and better interaction. The maximum elongation at break (7.1%) is observed for composite with 30 phr ELO because of plasticized epoxy matrix by unreacted long flexible chains of triglycerides. EPECO30SF composites exhibited lower strain at the break with respect to its ELO-based composite counterpart. This finding is well supported from relatively increased modulus/stiffness of EPECO30SF composite. Relatively higher stiffness and lower elongation of EPECO30SF composite reveal higher crosslinking through etherification caused by hydroxyl groups of ricinoleic acid. Despite lower epoxy content in ECO, almost all the epoxidized glyceride chains are reactive with abundant hydroxyl groups which can take part in the crosslinking process raising the brittleness of the composite. In contrast to this, ELO only contains epoxides which can undergo polymerization to contribute stiffness. At higher loading of ELO, the unreacted or poorly reactive glyceride chains plasticizes the matrix significantly which resulted in higher elongation and reduced modulus of the composites. However, inferior tensile strength was noticed for EPECOSF composites due to lower epoxy value of ECO and higher amount of ricinoleic acid content.

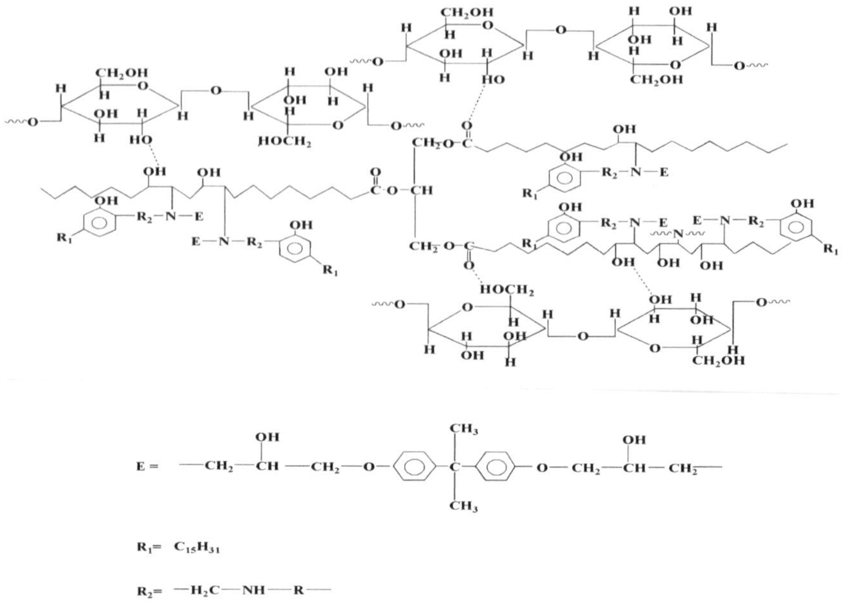

Fig. 8 Chemical interaction of bio-resin blend with cellulosic sisal fiber [102]

The probable chemical interaction of crosslinked epoxidized oil with cellulose moiety of sisal fibers is provided in Fig. 8. The –OH groups of sisal fibers attach with the –COO and –OH moieties of the cured bio-resin through hydrogen bonding resulting in superior strength and modulus. Additionally, –OH groups produced through ring opening of epoxy have the ability to form bonds with bio-fibers. Analogous interaction between cellulosic fibers and epoxidized oil (EO) modified epoxy with mechanism has been studied previously by Sahoo et al. [93]. It is noteworthy that the strength and modulus of EPELO20SF composites are maximum among all the samples and found to be exceeding than that of other eco-friendly EcoPoxy and Greenpoxy (>55% renewable content) composites with ∼23 vol.% flax fabric [90]. Particularly, EPELO20SF showed highest enhancement in strength and elastic modulus to the tune of 24 and 131%, correspondingly as compared to EcoPoxy/flax composite and nearly 6 and 91.6% increase, respectively compared to Greenpoxy/flax composite [90]. Similarly, EPECO20SF and EPECO30SF composites also showed acceptable stiffness and toughened properties suitable enough for specific applications.

6.4.3 Impact or Toughening Properties

Toughening property is very useful for construction and structural applications as it refers to energy absorbing ability of the specimen. To investigate the equivalent property, the influence of ELO and ECO on impact strength of the EPSF

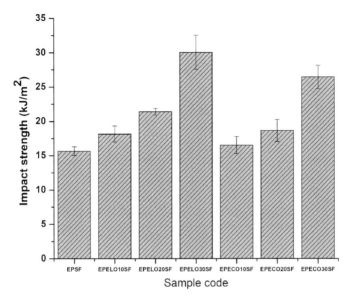

Fig. 9 Impact strength of epoxy bio-composites

composites has been depicted in Fig. 9. In case of composites, the nature of the resin, fiber and the resin–fiber interface decides the impact property.

The flexibility or ductility of the polymer matrix induced by epoxidized oil addition plays a vital role in raising the toughening property. The impact toughness raised with an increase in both ELO and ECO content which may be associated with an apparently increased strain at break. Both the bio-resins form random copolymers with epoxy through curing with PKA. Furthermore, epoxy resin reacts with carboxylic acid or hydroxy groups present as the ricinoleic acid moieties of ECO forming a relatively rigid crosslinked network [64, 104]. Due to this additional crosslinking of EPECO which resulted in increased rigidity, the impact strength of EPECOSF composites is found to be relatively inferior than that of EPELOSF counterparts. Addition of less viscous ELO into petro-based epoxy improved the impregnation of sisal fibers resulted in better packing and matrix–fiber physical interaction compared to viscous ECO. While a relatively higher rigidity of EPECO blend matrix offers lower impact strength to the composites, still a useful stiffness–toughness balance is well maintained.

6.4.4 Rheological Behavior of Bio-based Epoxy Resins

The variation of viscosity of uncured EP, ELO, ECO, EPELO20, and EPECO20 was studied and is illustrated in Fig. 10. All the resin except ELO exhibit shear thinning behavior. The viscosity of ELO and ECO at zero shear is measured to be 0.67 and 6.13 Pa.s, respectively. ELO displayed completely Newtonian flow

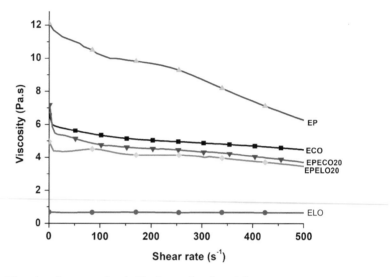

Fig. 10 Viscosity of epoxy and resin blends as a function of shear rate

behavior and constant viscosity regardless of shear rate chosen. Conversely, ECO displayed reduction in viscosity at lower shear rate and then became almost saturated at higher shear rate. Addition of 20 phr ELO and 20 phr ECO reduced the zero shear viscosity of EP from 11.45 Pa.s to 4.37 Pa.s and 7.19 Pa.s, respectively. It revealed that ELO lowers the viscosity of neat epoxy significantly demonstrating itself as a successful diluent. The short chain structure, less branching, and less entanglement of ELO decreased the viscosity of the blend to a higher order. The difference in flow behavior of the samples is observed due to structural and textural changes arising in entanglement during molecular alignment. ECO exhibits higher viscosity than ELO because of higher molecular weight, more entanglements, abundant hydroxyl groups, and larger intermolecular forces. However, both EPELO20 and EPECO20 blends display non-Newtonian flow behavior at low shear value followed by a transition into a Newtonian behavior which is desirable for some specific molding technique.

6.4.5   Effect of EO on Viscoelastic Behavior of Composites

To study the influence of ELO and ECO on sisal fiber reinforced epoxy composites, the thermophysical parameters like storage modulus (E') and loss factor (tan δ) of all the composites are measured and shown in Figs. 11 and Fig. 12.

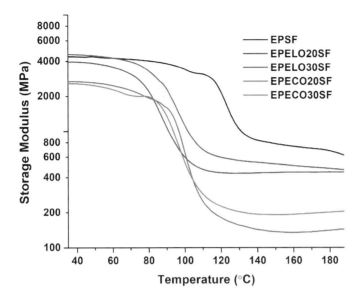

Fig. 11 Variation of storage modulus of bio-composites versus temperature

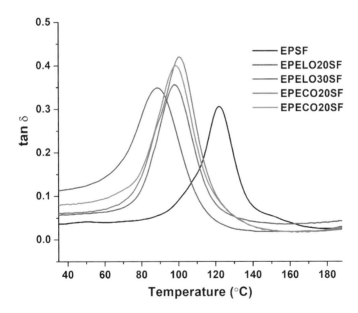

Fig. 12 Loss tangent curve of bio-composites versus temperature

Storage Modulus

In energy elastic region, the mobility of polymer chains is restricted enough to behave as rigid/stiff materials. The moduli are higher in all the samples in this glassy region and as the temperature is raised, these values drop rapidly and rubbery region starts exhibiting segmental motion. The modulus (E') of EPSF, EPELO20SF, and EPELO30SF at 35 °C were measured to be 4.37, 4.57, and 3.93 GPa, respectively and that of EPECO20SF and EPECO30SF were observed to be 2.67 and 2.57 GPa in that order. Higher E' of EPELO20SF may be ascribed to stronger fiber–matrix adhesion interface and better packing of fibers caused by relatively less viscous ELO bio-resin. Further, the intermolecular interactions between epoxidized oil and sisal fibers also offer rigidity or stiffness to the composites. The inferior modulus of EPECOSF composites may be attributed to the poor wetting and substandard fiber–matrix interface in the glassy region. However, in the rubbery state of the curve, E' of pure epoxy composite is found to be higher than all the bio-resin modified composites because of the presence of rigid aromatic groups [71]. Similar observations were noticed for ESO based jute and sisal fabric reinforced epoxy composites [103]. Likewise, 25% ESO-modified epoxy composite showed higher E' in glassy area as seen by Niedermann et al. [101]. On similar lines, Sahoo et al. reported an increase in modulus for sisal fiber reinforced EP/20% ESO composite [90].

Loss Factor (Tan δ)

In order to investigate the damping ability of bio-based composites, the loss tangent curve of all the composite systems is depicted in Fig. 12.

The curve showed broadening as the ELO was added to the epoxy composite and the degree of broadening raised with increase in ELO content from 20 to 30 phr. It is concluded that 30 phr ELO over plasticized the epoxy matrix and helped dissipate the higher amount of energy. Similarly, the peak intensity of tan δ gradually increased as ECO loading into the composites rose. It revealed the potential of the bio-based epoxy composites for energy absorbing ability, reduced noise and vibration regulating capability. The peak of tan δ curve of the neat EPSF composites goes toward left with an increase in EO content for both the systems (Table 6). This reduction in T_g may be explained due to decreased crosslink density

Table 6 Viscoelastic parameters of composites

Sample	E' at 35 °C (GPa)	E' at T_g (GPa)	Glass transition temperature T_g (°C)
EPSF	4.37	1.66	121
EPELO20SF	4.57	1.38	97
EPELO30SF	3.93	1.25	88
EPECO20SF	2.67	0.63	100
EPECO30SF	2.57	0.61	98

and undesirable plasticization caused by long flexible chains of ELO and ECO [101]. Still, the obtained modulus, loss factor and T_g of all modified composites are acceptable for specific applications.

6.4.6 Morphology of Fractured Surfaces

The micrographs of cracked surfaces of composites depicted in Fig. 13 are used to investigate the interfacial adhesion and failure.

No phase separation was noticed which indicates better miscibility of epoxidized oil with epoxy resin as well as with PKA crosslinker. For a composite to exhibit higher performance, good interfacial adhesion is needed between the resin and fiber after crosslinking. In case of EPSF composite (Fig. 13a), debonding at the interface, pull-out of fibers, and interfacial gap is observed which reflects the inability of composite in transferring load from the matrix to the fibers effectively resulting in inferior mechanical and thermomechanical properties than predicted. On the contrary, in EPELO30SF and EPECO30SF composites (Fig. 13b and c), better adhesion between sisal fiber mat and modified epoxy matrix is noticed without any interfacial gap resulting in easy stress transfer from matrix to the fibers. The fibers are found to be well-adhered to the matrix and no crack is noticed in matrix. This adhesion is seen due to proper wettability of fibers with less viscous epoxy/ELO or epoxy/ECO resin blend. A similar morphology which can contribute to higher mechanical performance was noticed for epoxidized soybean oil (ESO) and epoxidized hemp oil (EHO) based epoxy composites [123]. From this morphological study, it is revealed that addition of ELO and ECO improved the matrix–fiber adhesion and interaction.

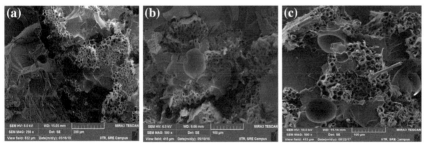

Fig. 13 FE-SEM micrographs of fractured surfaces of **a** EPSF, **b** EPELO30SF, **c** EPECO30SF bio-composites

7 Conclusions

The current chapter reviews the benefits and applications of plant oil based bio-resins blends and composites through an extensive literature survey of recently published articles. More specifically, the presented case study on ELO and ECO show a great deal of promise regarding the use of bio-resins as reactive diluents and lignocellulosic fibers as reinforcement within epoxy matrix to develop bio-composites for structural applications. ELO and ECO-based epoxy bio-composites reinforced with sisal fibers were prepared with large amount of bio-sourced content (60%). The tensile strength and elastic moduli of composites were best observed with 20 phr of ELO bio-resin addition due to proper impregnation of fibers and stronger matrix–fiber adhesion/interaction. Dynamic mechanical finding revealed raised storage modulus and damping factor with addition of epoxidized oil and sisal fibers into epoxy cured with bio-renewable crosslinker Better damping ability makes them a promising material for good shock absorption and energy dissipation under high vibration conditions. The incorporation of less viscous epoxidized oils enhanced the processibility, ensured better packing of fibers with no debonding or pull-out of fibers and improved stiffness–toughening balance Thus, ELO and ECO can be used as a potentially renewable reactive diluent to enhance the mechanical and thermomechanical properties of fiber reinforced composites for automotive and structural applications. Still, there are a lot of challenges, like lower thermal stability of natural fibers and plant oil based bio-resins, inferior stiffness, and undesired ductility of crosslinked bio-resins, etc. to be addressed for a wide use of plant oil based bio-composites. In this context renewable resourced nanofillers like nanocellulose, nanofibrils, nanostructured lignin, etc., can be used as reinforcing agents and bio-based aromatic epoxy monomers as a base matrix to develop composites with higher stiffness and stability.

Acknowledgements Science and Engineering Research Board (SERB), Government of India is highly acknowledged for NPDF funding support (File number: PDF/2015/000705).

References

1. Wool RP, Sun XS (2005) Bio-based polymers and composites. Elsevier Science and Technology Book
2. Gandini A (2008) Polymers from renewable resources: a challenge for the future of macromolecular materials. Macromolecules 41(24):9491–9504
3. Raquez JM, Deléglise M, Lacrampe MF, Krawczak P (2010) Thermosetting bio-materials derived from renewable resources. A critical review. Prog Polym Sci 35:487–509
4. Suttie E (2012) Bio-resins in construction: a review of current and future developments BRE publications, UK
5. Holbery J, Houston D (2006) Natural-fiber-reinforced polymer composites in automotive applications, JOM 58(11):80–86

6. Auvergne R, Caillol S, David G, Boutevin B, Pascault JP (2014) Biobased thermosetting epoxy: present and future. Chem Rev 114(2):1082–1115
7. Guner FS, Yagci Y, Erciyes AT (2006) Polymers from triglyceride oils. Prog Polym Sci 31 (7):633–670
8. Meier MAR, Metzger JO, Schubert US (2007) Plant oil renewable resources as green alternatives in polymer science. Chem Soc Rev 36(11):1788–1802
9. Lu Y, Larock RC (2009) Novel polymeric materials from vegetable oils and vinyl monomers: preparation, properties, and applications. Chemsuschem 2(2):136–147
10. Mercangoz M, Kusefoglu S, Akman U, Hortacsu O (2004) Polymerization of soybean oil via permanganate oxidation with sub/supercirifical CO_2. Chem Eng Process 43:1015–1027
11. Lligadas G, Ronda JC, Galia M, Cadiz V (2013) Renewable polymeric materials from vegetable oils. Mater Today 16:337–343
12. Ronda JC, Lligadas G, Galia M, Cadiz V (2011) Vegetable oils as platform chemicals for polymer synthesis. Eur J Lipid Sci Technol 113(1):46–58
13. Ronda JC, Lligadas G, Galia M, Cadiz V (2013) A renewable approach to thermosetting resins. React Funct Polym 7:381–395
14. Mittal V (2012) Renewable polymers: synthesis, processing and technology. John Wiley and Sons, Scrivener Publishing LLC
15. Xia Y, Larock RC (2010) Vegetable oil-based polymeric materials: synthesis, properties, and applications. Green Chem 12:1893–1909
16. Mallegol J, Lemaire J, Gardette JL (2000) Drier influence on the curing of linseed oil. Prog Org Coat 39(2–4):107–113
17. Guler OK, Guner FS, Erciyes AT (2004) Some empirical equations for oxypolymerization of linseed oil. Prog Org Coat 51:365–371
18. Keles E, Hazer B (2008) Autooxidized polyunsaturated oils/oily acids: post-it Applications and Reactions with Fe (III) and adhesion properties. Macromol Symp 269:154–160
19. Robertson ML, Chang KH, Gramlich WM, Hillmyer MA (2010) Toughening of polylactide with polymerized soybean oil. Macromolecules 43:1807–1814
20. Andjelkovic DD, Valverde M, Henna P, Li FK, Larock RC (2005) Novel thermosets prepared by cationic copolymerization of various vegetable oils synthesis and their structure property relationships. Polymer 46:9674–9685
21. Li FK, Larock RC (2001) New soybean oil styrene divinylbenzene thermosetting copolymers. I. Synthesis and characterization. J Appl Polym Sci 80:658–670
22. Patil H, Wagmare J (2013) Catalyst for epoxidation of oils: a review. Discovery 3(7):10–14
23. Gerhard K, Johannes TPD (1999) Recent developments in the synthesis of fatty acid derivatives. AOCS Press USA, p 157
24. Warwel S, Klaas MRG (1995) Chemo-enzymatic epoxidation of unsaturated carboxylic acids. J Molecular Catalysis B: Enzymatic 1:29–35
25. Okieimen FE, Bakare OI, Okieimen CO (2002) Studies on the epoxidation of rubber seed oil. Ind Crop Prod 15(2):139–144
26. Goud VV, Patwardhan AV, Pradhan NC (2006) Studies on the epoxidation of mahua oil *Madhumica indica* by hydrogen peroxide. Bioresource Technol 97:1365–1371
27. Goud VV, Pradhan NC, Patwardhan AV (2006) Epoxidation of karanja pongamia glabra oil by hydrogen peroxide. J Am Oil Chem Soc 83:635–640
28. Dinda S, Patwardhan AV, Goud VV, Pradhan NC (2008) Epoxidation of cottonseed oil by aqueous hydrogen peroxide catalysed by liquid inorganic acids. Bioresource Technol 99:3737–3744
29. Meyer PP, Techaphattana N, Manundawee S, Sangkeaw S, Junlakan W, Tongurai C (2008) Epoxidation of soybean oil and jatropha oil. Thammasat Int J Sci Tech 13:1–5
30. Cai C, Dai H, Chen R, Su C, Xu X, Zhang S, Yang L (2008) J Lipid Sci Technol 110:341–346
31. Petrivic ZS, Zlatanic A, Lava CC, Sinadinovic-Fiser S (2002) Epoxidation of soybean oil in toluene with peroxoacetic and peroxoformic acids-kinetics and side reactions. Euro J Lipid Sci Technol 104:293–299

32. Goud VV, Patwardhan AV, Dinda S, Pradhan NC (2007) Epoxidation of karanja (*Pongamia glabra*) oil catalysed by acidic ion exchange resin. Europ J Lipid Sci Technol 109:575–584
33. Mungroo R, Pradhan NC, Goud VV, Dalai AK (2008) Epoxidation of canola oil with hydrogen peroxide catalyzed by acidic ion exchange resin. J Am Oil Chem Soc 85:887–896
34. Sinadinovic F, Jankovic M, Petrovic ZS (2001) Kinetics of in-situ epoxidation of soybean oil in bulk catalysed by ion exchange resin. J Am Oil Chem Soc 78(7):725–731
35. Parada Hernandez NL, Bonon AJ, Bah JO, Barbosa MIR, Wolf Maciel MR, Filho RM (2016) Epoxy monomers obtained from castor oil using a toxicity-free catalytic system J Mol Catal A Chem
36. Dinda S, Goud V, Patwardhan AV, Pradhan NC (2011) Selective epoxidation of natural triglycerides using acidic ion exchange resin as catalyst. Asia-Pacific J Chem Engg 466
37. Pan X, Sengupta P, Webster DC (2011) Novel biobased epoxy compounds: epoxidized sucrose esters of fatty acids. Green Chem 13:965–975
38. Kim JR, Sharma S (2012) The development and comparison of bio-thermoset plastics from epoxidized plant oils. Ind Crop Prod 36:485–499
39. Zhu J, Chandrashekhara K, Flanigan V, Kapila S (2004) Curing and mechanical characterization of a soy-based epoxy resin system. J Appl Polym Sci 91:3513–3518
40. Holser RA (2008) Transesterification of epoxidized soybean oil to prepare epoxy methyl esters. Ind Crop Prod 27:130–132
41. Martini DS, Braga BA, Samios D (2009) On the curing of linseed oil epoxidized methyl esters with different cyclic dicarboxylic anhydrides. Polymer 50:2919–2925
42. Reiznautt QB, Garcia ITS, Samios D (2009) Oligoesters and polyesters produced by the curing of sunflower oil epoxidized biodiesel with cis-cyclohexane dicarboxylic anhydride. Synthesis and characterization. Mater Sci Eng C 29:2302–2311
43. Nicolau A, Mariath RM, Martini EA, Martini DS, Samios D (2010) The polymerization products of epoxidized oleic acid and epoxidized methyl oleate with cis-1,2 cyclohexanedicarboxylic anhydride and triethylamine as the initiator: Chemical structures, thermal and electrical properties. Mater Sci Eng C 30:951–962
44. Mustata F, Nita T, Bicu I (2014) The curing reaction of epoxidized methyl esters of corn oil with Diels-Alder adducts of resin acids. The kinetic study and thermal characterization of crosslinked products. J Anal Appl Pyrolysis 108:254–264
45. Wang R, Schuman TP (2013) Vegetable oil-derived epoxy monomers and polymer blends: a comparative study with review. eXP Polym Lett 7:272–292
46. Behera D, Banthia AK (2008) Synthesis, characterization, and kinetics Study of thermal decomposition of epoxidized soybean oil acrylate. J Appl Polym Sci 109:2583–2590
47. Fu L, Yang L, Dai C, Zhao C, Ma L (2010) Thermal and mechanical properties of acrylated expoxidized-soybean oil-based thermosets. J Appl Polym Sci 117:2220–2225
48. Li Y, Fu L, Lai S, Cai X, Yang L (2010) Synthesis and characterization of cast resin based on different saturation epoxidized soybean oil Eur. J Lipid Sci Technol 112:511–516
49. Rengasamy S, Mannari V (2013) Development of soy-based UV-curable acrylate oligomers and study of their film properties. Prog Org Coatings 76:78–85
50. Zhang P, Zhang J (2013) One-step acrylation of soybean oil (SO) for the preparation of SO-based macromonomers. Green Chem 15:641
51. Esen H, Küsefoğlu S, Wool R (2007) Photolytic and free-radical polymerization of monomethyl maleate esters of epoxidized plant oil triglycerides. J Appl Polym Sci 103(1):626–633
52. Esen H, Küsefoğlu SH (2003) Photolytic and free-radical polymerization of cinnamate esters of epoxidized plant oil triglycerides. J Appl Polym Sci 89(14):3882–3888
53. Can E, Küsefoğlu S, Wool RP (2001) Rigid, thermosetting liquid molding resins from renewable resources. I. Synthesis and polymerization of soy oil monoglyceride maleates. J Appl Polym Sci 81(1):69–77
54. Wang CS, Yang LT, Ni BL, Shi G (2009) Polyurethane networks from different soy-based polyols by the ring opening of epoxidized soybean oil with methanol, glycol, and 1,2-propanediol. J Appl Polym Sci 114(1):125–131

55. Dai H, Yang LT, Lin B, Wang CS, Shi G (2009) Synthesis and characterization of the different soy-based polyols by ring opening of epoxidized soybean oil with methanol, 1,2-ethanediol and 1,2-propanediol. J Am Oil Chem Soc 86(3):261–267
56. Sahoo SK, Khandelwal V, Manik G (2018) Development of completely bio-based epoxy networks derived from epoxidized linseed and castor oil cured with citric acid. Polym Adv Technol. https://doi.org/10.1002/pat.4316
57. Yim YJ, Rhee KY, Park SJ (2017) Fracture toughness and ductile characteristics of diglycidyl ether of bisphenol-A resins modified with biodegradable epoxidized linseed oil, Compos Part B Eng 131:144–152
58. Espana JM, Nacher LS, Boronat T, Fombuena V, Balart R (2012) Properties of biobased epoxy resins from epoxidized soybean oil (ESBO) cured with maleic anhydride (MA). J Am Oil Chem Soc 89(11):2067–2075
59. Liu Z, Erhan SZ (2010) Ring-opening polymerization of epoxidized soybean oil. J Am Oil Chem Soc 87(4):437–444
60. Park SJ, Jin FL, Lee JR (2004) Synthesis and thermal properties of epoxidized vegetable oil. Macromol Rapid Commun 25(6):724–727
61. Jin FL, Park SJ (2007) Thermal and rheological properties of vegetable oil-based epoxy resins cured with thermally latent initiator. J Ind Eng Chem 13(5):808–814
62. Unnikrishnan KP, Thachil ET (2006) Toughening of epoxy resins. Des Monomers Polym 9 (2):129–152
63. Thomas R, Yumei D, Yuelong H, Moldenaers Y, Le P, Weimin Y, Czigany T, Thomas S (2008) Miscibility, morphology, thermal, and mechanical properties of a DGEBA based epoxy resin toughened with a liquid rubber. Polymer 49:278–294
64. Zhang C, Xia Y, Chen R, Huh S, Johnston P, Kessler MR (2013) Soy-castor oil based polyols prepared using a solvent-free and catalyst-free method and polyurethanes therefrom. Green Chem 15(6):1477
65. Jin FL, Park SJ (2008) Thermomechanical behavior of epoxy resins with epoxidized vegetable oil. Polym Int 57:577–583
66. Tan SG, Chow WS (2010) Thermal properties, fracture toughness and water absorption of epoxy-palm oil blends. Polymer-Plastics Technol Eng 49(9):900–907
67. Tan SG, Chow WS (2010) Biobased epoxidized vegetable oils and its greener epoxy blends: a review. Polym Plast Technol Eng 49:900–907
68. Sahoo SK, Mohanty S, Nayak SK (2015) Synthesis and characterization of bio-based epoxy blends from renewable resource based epoxidized soybean oil as reactive diluent. Chi J Polym Sci 33(1):137–152
69. Debnath D, Khatua BB (2011) Preparation by suspension polymerization and characterization of polystyrene (PS)-poly(methyl methacrylate) (PMMA) core-shell nanocomposites. Macromol Res 19(6):519–527
70. Ganan P, Garbizu S, Ponte RL, Mondragon I (2005) Surface modification of sisal fibers: effects on the mechanical and thermal properties of their epoxy composites. Polym Composites 26(2):121–127
71. Sahoo SK, Mohanty S, Nayak SK (2015) Toughened bio-based epoxy blend network modified with transesterified epoxidized soybean oil: synthesis and characterization. RSC Adv 5:13674–13691. https://doi.org/10.1039/C4RA11965G
72. Muturi P, Wang D, Dirlikov S (1994) Epoxidized vegetable oils as reactive diluents I. Comparison of vernonia, epoxidized soybean and epoxidized linseed oils. Prog Org Coat 25:85–94
73. Ratna D (2001) Mechanical properties and morphology of epoxidized soybean-oil-modified epoxy resin. Polym Int 50:179–184
74. Park SJ, Jin FL, Lee JR (2004) Thermal and mechanical properties of tetrafunctional epoxy resin toughened with epoxidized soybean oil. Mater Sci Eng A 374(1–2):109–114
75. Park SJ, Jin FL, Lee JR (2004) Effect of biodegradable epoxidized castor oil on physicochemical and mechanical properties of epoxy resins. Macromol Chem Phys 205:2048–205

76. Jin FL, Park SJ (2008) Impact-strength improvement of epoxy resins reinforced with a biodegradable polymer. Mater Sci Eng A 478(1–2):402–405
77. Miyagawa H, Mohanty AK, Misra M, Drzal LT (2004) Thermo-physical and impact properties of epoxy containing epoxidized linseed oil 1. Macromol Mater Eng 289(7):629–635
78. Miyagawa H, Mohanty AK, Misra M, Drzal LT (2004) Thermo-physical and impact properties of epoxy containing epoxidized linseed oil 2. Macromol Mater Eng 289(7):636–641
79. Miyagawa H, Misra M, Drzal LT, Mohanty AK (2005) Fracture toughness and impact strength of anhydride-cured biobased epoxy. Polym Eng Sci 45:487–495
80. Tan SG, Chow WS (2010) Thermal properties of anhydride-cured bio-based epoxy blends J Therm Anal Calorim 101(3):1051–1058
81. Altuna FI, Esposito LH, Ruseckaite RA, Stefani PM (2011) Thermal and mechanical properties of anhydride-cured epoxy resins with different contents of biobased epoxidized soybean oil. J Appl Polym Sci 120:789–798
82. Altuna FI, Pettarin V, Martin L, Retegi A, Mondragon I, Ruseckaite RA, Stefani PM (2014) Copolymers based on epoxidized soy bean oil and diglycidyl ether of bisphenol A: relation between morphology and fracture behavior polym. Eng Sci 54:569
83. Chen Y, Yang L, Wu J, Ma L, Finlow DE, Song SLK (2013) Thermal and mechanical properties of epoxy resin toughened with epoxidized soybean oil. J Therm Anal Calorim 11. (2):939–945
84. Mustata F, Tudorachi N, Rosu D (2011) Curing and thermal behavior of resin matrix for composites based on epoxidized soybean oil/diglycidyl ether of bisphenol A. Composite Part B 42(7):1803–1812
85. Mohanty AK, Misra M, Drzal LT (2005) Natural fibers, biopolymers, and biocomposites [chapter 2]. CRC Press, Taylor & Francis Group, New York
86. O'Donnell A, Dweib MA, Wool RP (2004) Natural fiber composites with plant oil-based resin. Compos Sci Technol 64:1135–1145
87. Liu Z, Erhan SZ, Barton FE, Akin DE (2006) Green composites from renewable resources preparation of epoxidized soybean oil and flax fiber composites. J Agric Food Chem 54:2134–2137
88. Tran P, Graiver D, Narayan R (2006) Biocomposites synthesized from chemically modified soy oil and biofibers. J Appl Polym Sci 102:69–75
89. Boquillon N (2006) Use of an epoxidized oil-based resin as matrix in vegetable fibers-reinforced composites. J Appl Polym Sci 101:4037–4043
90. Lee KY, Ho KKC, Schlufter K, Bismarck A (2012) Hierarchical composites reinforced with robust short sisal fibre performs utilising bacterial cellulose as binder. Compos Sci Technol 72:1479–1486
91. Bertomeu D, Sanoguerab DG, Fenollar O, Boronat T, Balart R (2012) Polym Compos 33:683
92. Zhu J, Chandrashekhara K, Flanigan V, Kapila S (2004) Manufacturing and mechanical properties of soy-based composites using pultrusion. Compos A 35:95–101
93. Sahoo SK, Mohanty S, Nayak SK (2015) Study on the effect of woven sisal fiber mat on mechanical and viscoelastic properties of petroleum based epoxy and bioresin modified toughened epoxy network. J Appl Polym Sci 42699:n/a-n/a. https://doi.org/10.1002/app42699
94. Faruk O, Bledzki AK, Fink HP, Sain M (2012) Biocomposites reinforced with natural fibers: 2000–2010. Prog Polym Sci 37:1552–1596
95. Harris B (1999) Engineering composite materials. The Cambridge University Press, London
96. Joshi SV, Drzal LT, Mohanty AK, Arora S (2004) Are natural fiber composites environmentally superior to glass fiber reinforced composites? Compos A 35:371–376
97. Thakur VK, Thakur MK (2014) Processing and characterization of natural cellulose fibers/ thermoset polymer composites. Carbohyd Polymer 109:102–117

98. Satyanarayana KG, Arizaga GGC, Wypych F (2009) Biodegradable composites based on lignocellulosic fibers-an overview. Prog Polym Sci 34:982–1021
99. Mallick PK (2007) Fiber reinforced composites; materials, manufacturing and design. Taylor & Francis, New York
100. Mosiewicki MA, Aranguren MI (2013) A short review on novel biocomposites based on plant oil precursors. Eur Polym J 49:1243–1256
101. Niedermann P, Szebenyi G, Toldy A (2017) Effect of epoxidized soybean oil on mechanical properties of woven jute fabric reinforced aromatic and aliphatic epoxy resin composites. Polym Compos 38:884–892. https://onlinelibrary.wiley.com/doi/abs/10.1002/pc.23650
102. Sahoo SK, Khandelwal V, Manik G (2018) Influence of epoxidized linseed oil and sisal fibers on structure–property relationship of epoxy biocomposite. Polym Compos. https://doi.org/10.1002/pc.24857
103. Judith D, Espinoza P, Brent AN, Darrin MH, Zhigang CCA, Ulven DPW (2011) Comparison of curing agents for epoxidized vegetable oils applied to composites. Polym Compos 32(11):1806–1816
104. Kunduru KR, Basu A, Haim Zada M, Domb AJ (2015) Castor oil-based biodegradable polyesters. Biomacromolecules 16(9):2572–2587
105. Petrovic ZS (2010) Polymers from biological oils. Contemp Mater 1(1):39–50
106. Espinosa LMD, Meier MAR (2011) Plant oils: the perfect renewable resource for polymer science. Euro Polym J 47(5):837–852
107. Lu J, Khot S, Wool RP (2005) New sheet molding compound resins from soybean oil. I. Synthesis and characterization. Polymer 46:71–80
108. Zhu J, Chandrashekhara K, Flanigan V, Kapila S (2004) Curing and mechanical characterization of a soy-based epoxy resin system. J Appl Polym Sci 91:3513–3518
109. Shabeer A, Garg A, Sundararaman S, Chandrashekhara K, Flanigan V, Kapila S (2005) Dynamic mechanical characterization of a soy based epoxy resin system. J Appl Polym Sci 98(4):1772–1780
110. Shabeer A, Sundararaman S, Chandrashekhara K, Dharani LR (2007) Physicochemical properties and fracture behavior of soy-based resin. J Appl Polym Sci 105:656–663
111. Grishchuk S, Kocsis KJ (2012) Modification of vinyl ester and vinyl ester-urethane resin-based bulk molding compounds (BMC) with acrylated epoxidized soybean and linseed oils. J Mater Sci 47(7):3391–3399
112. Park SJ, Jin FL, Lee JR (2004) Effect of biodegradable epoxidized castor oil on physicochemical and mechanical properties of epoxy resins. Macromol Chem Phys 205(15):2048–2054
113. Haq M, Burgueño R, Mohanty AK, Misra M (2008) Hybrid bio-based composites from blends of unsaturated polyester and soybean oil reinforced with nanoclay and natural fibers. Compos Sci Technol 68:3344–3351
114. Ramamoorthy SK, Di Q, Adekunle K, Skrifvars M (2012) Effect of water absorption on mechanical properties of soybean oil thermosets reinforced with natural fibers. J Reinf Plastics Compos 31(18):1191–1200
115. Li Y, Mai YW, Ye L (2000) Sisal fibre and its composites: a review of recent developments. Compos Sci Technol 60:2037–2055
116. Sahoo SK, Khandelwal V, Manik G (2015) Effect of lignocellulosic fibers on mechanical, thermomechanical and hydrophilic studies of epoxy modified with novel bioresin epoxy methyl ester derived from soybean oil. Polym Adv Technol. https://doi.org/10.1002/pat.3592
117. Sahoo SK, Mohanty S, Nayak SK (2016) Dynamic mechanical and interfacial properties of sisal fiber reinforced composite with epoxidized soybean oil based epoxy matrix. Polym Compos. https://doi.org/10.1002/pc.24168
118. Badrinath R, Senthilvelan T (2014) Comparative investigation on mechanical properties of banana and sisal reinforced polymer based composite. Procedia Mater Sci 5:2263–2272
119. Hodzic A, Shanks R (2014) Wood head natural fibre composites materials, processes and applications publishing limited [chapter 9], pp 233–270

120. Åkesson D, Skrifvars M, Walkenstrom P (2009) Preparation of thermoset composites from natural fibres and acrylate modified soybean oil resins. J Appl Polym Sci 114:2502–2508

121. Adekunle K, Cho SW, Ketzscher R, Skrifvars M (2012) Mechanical properties of natural fiber hybrid composites based on renewable thermoset resins derived from soybean oil fo use in technical applications. J Appl Polym Sci 124:4530–4541

122. Adekunle K, Patzelt C, Kalantar A, Skrifvars M (2011) Mechanical and viscoelastic properties of soybean oil thermoset reinforced with jute fabrics and carded lyocell fiber J Appl Polym Sci 2855–2863

123. Manthey NW, Cardona F, Francucci G, Aravinthan T (2013) Thermo-mechanical properties of epoxidized hemp oil-based bioresins and biocomposites. J Reinf Plast Compos 32:1444–1456

124. Fejos M, Karger KJ, Grishchuk S (2013) Effects of fibre content and textile structure on dynamic-mechanical and shape-memory properties of ELO/flax biocomposites. J Reinf Plas Compos 32(24):1879–1886

125. Marrot L, Bourmaud A, Bono P, Baley C (2014) Multi-scale study of the adhesion between flax fibers and biobased thermoset matrices. Mater Design 62:47–56

126. Rana A, Evitts RW (2015) Development and characterization of flax fiber reinforced biocomposite using flaxseed oil-based bio-resin. J Appl Polym Sci 132:41807

127. Ding C, Matharu AS (2014) Recent developments on biobased curing agents: a review o their preparation and use. ACS Sustain Chem Eng 2:2217–2236. https://doi.org/10.1021. sc500478f

128. Sahoo SK, Khandelwal V, Manik G (2017) Development of toughened bio-based epoxy with epoxidized linseed oil as reactive diluent and cured with bio-renewable crosslinker Polym Adv Technol:1–10. https://doi.org/10.1002/pat.4166

129. Pathak SK, Rao BS (2006) Structural effect of phenalkamines on adhesive viscoelastic and thermal properties of epoxy networks. J Appl Polym Sci 102:4741–4748. https://doi.org/10 1002/app.25005

130. Voirin C, Caillol S, Sadavarte NV, Tawade BV, Boutevin B, Wadgaonkar PP (2014 Functionalization of cardanol: towards biobased polymers and additives. Polym Chem 5:3142–3162. https://doi.org/10.1039/c3py01194a

131. Khandelwal V, Sahoo SK, Kumar A, Manik G (2018) Electrically conductive green composites based on epoxidized linseed oil and polyaniline: an insight into electrical, thermal and mechanical properties. Compos Part B Eng 136:149–157

132. Sacristan M, Ronda JC, Galia Ma, Cadiz V (2010) Synthesis and properties of boron-containing soybean oil based thermosetting copolymers. Polymer 51:6099–6106

133. Prasad AVR, Rao KM (2011) Mechanical properties of natural fibre reinforced polyester composites: Jowar sisal and bamboo. Mater Design 32(8–9):4658–4663

Printed in the United States
By Bookmasters